NEURAL NETWORKS

Prentice Hall Advanced Reference Series

Engineering

ANTOGNETTI AND MILUTINOVIC, EDS. *Neural Networks: Concepts, Applications, and Implementations, Volumes I, II, III, and IV*

DENNO *Power System Design and Applications for Alternative Energy Sources*

ESOGBUE, ED. *Dynamic Programming for Optimal Water Resources Systems Analysis*

FERRY, AKERS, AND GREENEICH *Ultra Large Scale Integrated Microelectronics*

GNADT AND LAWLER, EDS. *Automating Electric Utility Distribution Systems: The Athens Automation and Control Experiment*

HALL *Biosensors*

HAYKIN, ED. *Advances in Spectrum Analysis and Array Processing, Volumes I and II*

HAYKIN, ED. *Selected Topics in Signal Processing*

HENSEL *Inverse Theory and Applications for Engineers*

JOHNSON *Lectures on Adaptive Parameter Estimation*

MILUTINOVIC, ED. *Microprocessor Design for GaAs Technology*

QUACKENBUSH, BARNWELL III, AND CLEMENTS *Objective Measures of Speech Quality*

ROFFEL, VERMEER, AND CHIN *Simulation and Implementation of Self-Tuning Controllers*

SASTRY AND BODSON *Adaptive Control: Stability, Convergence, and Robustness*

SWAMINATHAN AND MACRANDER *Materials Aspects of GaAs and InP Based Structures*

TOMBS *Biotechnology in the Food Industry*

NEURAL NETWORKS
Concepts, Applications, and Implementations
Volume II

Paolo Antognetti and Veljko Milutinović,
Editors

Prentice Hall, Englewood Cliffs, New Jersey 07632

Library of Congress Cataloging-in-Publication Data

Neural networks.
    (Prentice Hall advanced reference series. Engineering.
    Includes bibliographical references and index.
    1. Neural networks (Computer science) I. Antognetti,
Paolo. II. Milutinovic, Veljko. III. Series.
QA76.87.N48 1991   006.3   90-7485
ISBN 0-13-612516-6 (v. 1)
ISBN 0-13-612763-0 (v.2)

Cover Design: *Karen A. Stephens*
Manufacturing buyer: *Kelly Behr*
Acquisitions Editor: *Bernard Goodwin*

Prentice Hall Advanced Reference Series

© 1991 by Prentice-Hall, Inc.
A division of Simon & Schuster
Englewood Cliffs, New Jersey 07632

The publisher offers discounts on this book when ordered
in bulk quantities. For more information, write:

> Special Sales/College Marketing
> Prentice-Hall, Inc.
> College Technical and Reference Division
> Englewood Cliffs, New Jersey 07632

ISBN 0-13-612763-0

Printed in the United States of America
10 9 8 7 6 5 4 3 2 1

ISBN 0-13-612763-0

PRENTICE-HALL INTERNATIONAL (UK) LIMITED, *London*
PRENTICE-HALL OF AUSTRALIA PTY. LIMITED, *Sydney*
PRENTICE-HALL CANADA INC., *Toronto*
PRENTICE-HALL HISPANOAMERICANA, S.A., *Mexico*
PRENTICE-HALL OF INDIA PRIVATE LIMITED, *New Delhi*
PRENTICE-HALL OF JAPAN, INC., *Tokyo*
SIMON & SCHUSTER ASIA PTE. LTD., *Singapore*
EDITORA PRENTICE-HALL DO BRASIL, LTDA., *Rio de Janeiro*

Contents

Foreword

When interest in neural networks revived some fifteen years ago, few people believed that such systems would ever be of any use. Computers worked too well; it was felt that they could be programmed to perform any desired task.

Clearly, fashion has changed. Now, limitations of current computers in solving many problems involving difficult to define rules or complex pattern recognition are widely recognized; if anything, expectations for neural networks may be too high. The problem is no longer to convince anyone that neural networks might be useful, but rather to actually incorporate such networks into systems that solve real-world problems economically.

Neural networks are inspired by biological systems where large numbers of neurons, that individually function rather slowly and imperfectly, collectively perform tasks that even the largest computers have not been able to match. They are made of many relatively simple processors connected to one another by variable memory elements whose weights are adjusted by experience. They differ from the now standard Von Neumann computer in that they characteristically process information in a manner that is highly parallel rather than serial, and that they learn (memory element weights and thresholds are adjusted by experience) so that to a certain extent they can be said to program themselves. They differ from the usual artificial intelligence systems in that (since neural networks learn) the solution of real-world problems requires much less of the expensive and elaborate programming and knowledge engineering required for such artificial intelligence products as rule-based expert systems.

In their current state, neural networks are probably best at problems related to pattern recognition. Some existing neural network systems can efficiently and rapidly learn to separate enormously complex decision spaces. The problem of coordinating many neural networks, each a specialist in dividing some portion of the decision space, has also been solved. It is in these areas, therefore, that the first commercial uses will appear. Products that recognize characters, assembly line parts or signatures, that make complex decisions mimicking or improving on human experts (such as underwriters) that can diagnose engine or assembly line problems are in the prototype stage and/or are already fielded. One expects, further, that the pattern recognition ability coupled with, and feeding back and forth to rule-based systems (as has already been done in some simple applications) will finally result in machines that share our ability to learn and duplicate our processes of reasoning—machines that might be said to think.

The question is not whether but when.

Predicting the future, as we all know, is risky. Predicting the evolution of new technology is downright hazardous. Who in the 1930's would have said that among the consequences of the uncertainty principle would be transistors, silicon chips, and all of the vast array of solid state devices on which all modern computers depend? Or that superconductors would lead to extraordinarily sensitive detectors of magnetic fields now carried on many naval ships? Or in the late 19th century, that among the consequences of the research of Maxwell, Lorentz and Einstein, would be all that we call modern communication: radio, radar, etc.?

Accepting this risk, I would predict that neural networks will become standard components of what we today call computers. This will likely occur in a somewhat evolutionary manner: they will encroach gradually—board by board, intelligent components, that can be trained by humans in a language humans understand, into dumb machines—somewhat like neo-cortex came to dominate the reptilian brain. And, just as the 20th century is the century of automobiles, airplanes, telephones and computers, the 21st will be the century of intelligent machines. We will not only learn to live with these machines but, indeed, will wonder, one day, how we ever lived without them.

Leon Cooper
1972 Nobel Laureate

Preface

Neural networks and neural engineering in the wide sense are among the most rapidly developing scientific fields today. The field is interdisciplinary in its nature, and the success formula is to combine expertise and experience from several scientific fields.

The basic idea behind this series of books is to combine expertise and experience of contributing authors from a number of different scientific disciplines.

The style and structure of the books are oriented to those who need to get a broad insight into the state-of-the-art developments in the field. Each book in the series includes three parts: Concepts, Applications, and Implementations.

We would like to thank all contributing authors for their efforts and enthusiasm. Most of them are also the major contributors to the field itself, and we feel honored to have had an opportunity to work with them.

We would also like to thank Professor Leon Cooper for his creative role during our meetings in Genoa, Italy. His foreword to this series sheds an important new light on the future development of the field.

Finally, we would like to thank our colleagues from Genoa Research, Italy (especially Dr. Marenco, Mrs. Arata, and Ms. Ponta), and the University of Belgrade, Yugoslavia (especially Professors Dujmovic, Djordjevic, Lazic, Jovanovic, Oklobdzija, and Velasevic), for their suggestions and encouragement.

Please note that the Appendix provides a complete listing of contents and contributing authors for all four volumes of *Neural Networks*.

Paolo Antognetti and Veljko Milutinović, Editors

NEURAL NETWORKS

1

NEURAL NETWORKS:
A COMPUTATIONAL PERSPECTIVE

S. S. Iyengar and R. L. Kashyap

1.   Introduction

In the past few years the computational approach to artificial intelligence and cognitive engineering has undergone a significant evolution. Not only is this transformation, from discrete symbolic reasoning to massively parallel, connectionist neural modeling, of compelling scientific interest, but also is of considerable practical interest. It is changing the very rubric of information processing and problem solving. In general, the scientific and engineering community is contested with two basic categories of problems. Problems that are clearly defined and deterministic. They are targeted for situations that are completely deterministic, precisely controllable, and can best be handled by computers employing rigorous, precise logic, algorithms, or production rules. This class deals with *structured problems* such as data processing and bookkeeping, automated assembly in controlled workspace, etc. On the other hand, there are scenarios such as maintenance of nuclear plants, undersea mining, battle management, assembly/repair of space satellites, that lead to computational problems that are inherently ill-posed and ill-conditioned [90,97]. Such *unstructured problems* entail providing for situations that may have received no prior treatment or thought. Decisions need to be made, based on information that is incomplete, often ambiguous, plagued with imperfect or inexact knowledge, and involve the handling of large sets of competing constraints that can tolerate "close enough" solutions. The outcome depends, on very many inputs and their statistical variations, and there is not a clear logical method for arriving at the answer. In summary, this category encapsulates problems that cannot be satisfactorily addressed using traditional computational paradigms [3], such as Random Access Machines (RAM), Markov Algorithms, Universal Turing Machines, Production Systems, etc. The focus of Artificial Intelligence and machine learning has traditionally been to understand, and engineer systems that can address such unstructured computation problems.

However, it is observed that the biological machinery is capable of providing satisfactory solutions to such ill-structured problems with remarkable ease and flexibility [7,40,64,77]. The key thrust underlying any paradigmatic development today, is to understand how the above unstructured computations, are interpreted, organized and carried out by the biological systems. The latter exhibit a spontaneous emergent ability that enables them to adapt their structure and function. For example, a five-year old child can trivially recognize and distinguish among the different types of animals he sees, while similar perceptual interpretation is still a formidable task for even the most highly evolved microelectronic pattern recognition devices and rule based expert systems. Algorithmic mechanisms simply can't match the biological machinery when it comes to taking sensory information and acting on it, specially when the sensors are bombarded by a range of different, and in some cases competing images. Besides, the only unquestionably intelligent existing systems, are the nervous systems of biological organisms. Engineered intelligent systems, e.g., expert systems, exploratory autonomous rovers etc., behave with remarkable rigidity when compared to their biological counterparts, especially in their ability to recognize objects or speech, to manipulate and adapt in an nonstationary environment and to learn from past experience. They lack common sense knowledge and reasoning, have little sense of similarity or repetition or pattern - they do not know their own limitations. They are insensitive to context and are likely to give incorrect responses to queries that are outside the domains for which they are programmed.

A major reason for this lack of broad technical success in emulating some of the fundamental aspects of human intelligence, lies in the differences between the organization, and the dynamics of biological neuronal circuitry and its emulation using the symbolic processing paradigm[26,51]. For example, it is widely hypothesized [46,69-71,107] that analogy and reminding, guide all our thought patterns and that being attuned to vague resemblances is the hallmark of intelligence, it would be naive to expect that logical manipulation of symbolic descriptions as an adequate tool. Furthermore, there is psychophysical evidence [20,29,58] that while the beginner learns through rules, the expert discards such rules. Instead he discriminates thousands of patterns in his domains of expertise acquired through experience, and how to respond to them. Thus, it is rapidly becoming evident that many of the unstructured problems characterized above, can be best solved not with traditional AI techniques, but by "analogy", "subsymbolic" [95] or pattern matching techniques. While AI attempts to do this, neural networks, a biologically inspired, computational and information processing paradigm, provides us with an inherently better, but not the only tool. networks. In the sequel, we shall demonstrate in this paper that many of the unstructured problems characterized above can be best solved not with traditional AI techniques, but by analogy or pattern matching techniques. While AI attempts to do this, neural networks, a biologically motivated paradigm, provide us with an inherently better, but not the only, tool. Our main objective in this paper is to demonstrate the applicability of the phenomenology of neural networks, aimed at solving classes of unstructured problems at the core of AI thrust. In particular, we contrast neural networks with the symbolic computing paradigm, thereby emphasizing relaxation, and not heuristic search as the basis of achieving biological capability.

2. Artificial Neural Networks

Neurons are nerve cells and neural networks are networks of such cells. The cerebral cortex of the brain is an example of a natural neural network. Though the notion of imitating human neuronal processing system, cognitive capabilities and biological phenomena to model computation for AI research was introduced a couple of decades back, it is only recently that they are gaining recognition as a viable alternative to the traditional Von Neumann architectures. Recent advances in our understanding of anatomical and functional architecture [18,76] chemical composition [2], electrical and organizational processes [21,52,72,109] occurring in the brain and nervous system along with the advances in hardware technology and capability are leading to physical and electro-optical realizations [48] of randomly organized interconnect, networks with computationally useful collective properties such as time sequence retention, error correction and noise elimination, recognition, generalization, etc. As shown in figure 1, *Artificial Neural Systems* are characterized as distributed computational system comprising of a large number of processing units each of which has selected characteristics of biological neurons connected to each other in a directed graph of varying configuration [48]. They can be defined [63] as "massively parallel, adaptive dynamical systems modeled on the general features of biological networks, that can carry out useful information processing by means of their state response to initial or continuous input. Such neural systems interact with the objects of the real world and its statistical characteristics in the same way as biological systems do." In contrast to the existing notions on applicative and symbolic computing, the potential advantages of neuronal processing arise as a result of their ability to perform massively parallel, asynchronous and distributed information processing. Neurons with simple properties and interacting according to relatively simple rules can accomplish collectively complex functions. This is based on their ability to provide a collectively-computed solution to a problem on the basis of analog input information resulting from a high degree of random interconnectivity, storage and simplicity of individual operations. Neural network modeling then is a discipline which attempts to "understand brain- like computational systems" [10,60] and has been variously termed as "computational neuroscience" [7], "parallel distributed processing"

[95], "connectionism" [85], etc.

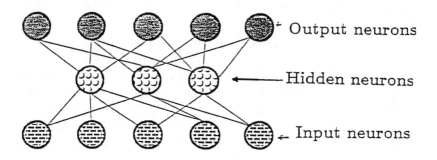

Fig. 1 Graph-theoretic representation of Artificial Neural Systems.

The bulk of neural network models can be classified into two categories; those that are intended as computational models of biological nervous systems or neurobiological models [21,56,75], and those that are intended as biologically-inspired models of computational devices with technological applications, also referred to as Artificial Neural Systems (ANS) [4,5,22,53,54]. The latter class demonstrates interesting capabilities which suggest that they might form useful components in intelligent systems. Though the primary emphasis of this paper is on ANS, we will be highlighting the influence of neurobiology on the formulation of artificial neural models, and the resulting computational implications. To get a sense of the required number and interconnectivity of neuronal circuitry for intelligent behavior, we begin by examining biological neural networks. Most existing neural network models are based on idealizations of the biological neuron and the synaptic conduction mechanisms, shown in figure 2(a) and 2(b), respectively. As shown in Fig 2(a), each neuron is characterized by a cell body or the cyton and thin branching extensions called dendrites and axons that are specialized for inter-neuron transmission. The dendrites receive inputs from other neurons and the axon provides outputs to other neurons. The neuron itself is imbedded in an aqueous solution of ions, and its selective permeability to these ions establishes a potential gradient responsible for transmitting information. Neurons receive electrochemical input signals from other neurons to which they are connected at sites on their surface, called synapses (see Fig. 2(b)), whose number varies from 10000 to 100000 for each neuron. The input signals are combined in various ways, triggering the generation of an output signal by a special region near the cell body. However, the neurobiological phenomenon that is of particular interest, is the changing chemistry of synapse as information flows from one neuron to another. On the transmitting or pre-synaptic side of the synapse, triggering of the synaptic pulse releases a neurotransmitter, that diffuses across a gap to the receiving side of the synapse. On the post-synaptic or receiving side, the neurotransmitter binds itself to receptor molecules, thereby affecting the ionic channels and changing the electrochemical potential. The magnitude of this change is determined by many factors local to the synapse, e.g., amount of neurotransmitter released, number of post-synaptic receptors, etc. Therefore, neurocomputation, biological self-organization, adaptive learning and other mental phenomena are largely manifested in changing the effectiveness or "strength" of the synapse and their topology. For details on the biological neuron, membrane polarization chemistry and synaptic modification the reader may refer to [9,100].

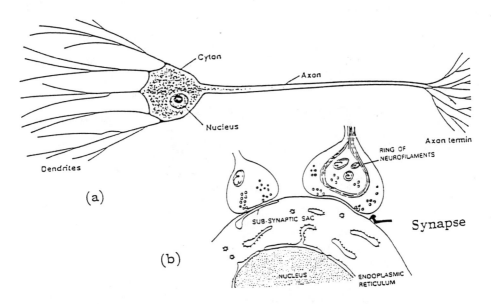

Figure 2:(a) Biological neuron, and (b) morphology of neuron-to-neuron connection or synapse. Adapted from Ito [56].

This understanding has led to the formulation of *simulated neurons*, a basic building block of neural network models, whose functional model is shown in figure 3. Four useful areas can be abstracted out. The first is the synapse where signals are passed from one neuron to another, and the amount of signal is regulated, i.e., gated or weighted by the strength of the synaptic interconnection. In the activated neuron region, denoted as *summer*, synaptic signals containing excitatory and inhibitory information are combined to affect the tendency of a cell to fire or stay put. The threshold detector determines if the neuron is actually going to fire or not, while axonal paths conduct the output activation energy to other synapses to which the neuron is connected. Of particular applicational and modeling interest, are the mathematical notions of synapse and synaptic modification, mechanisms by which such units can be connected together to compute, and the rules whereby such interconnected systems to could be made to learn. But before, we go into the mathematical characterization of neural models, we identify the basic attributes pertaining to neural networks.

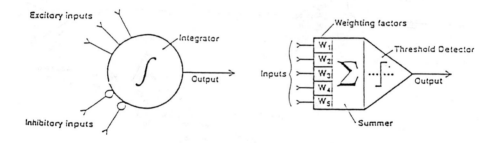

Fig. 3   Functional model of a simulated neuron.

3. Computational Characterization

Development of detailed mathematical models began with the work of McCulloch and Pitts [78], Minsky and Papert [81], Hebb [47], Rosenblatt [93], Widrow and Hoff [104]. However, the resurgence of the field is due to the more recent theoretical contributions by Kohonen [60,61,62], Grossberg [38,39,39], Amari [5,6], Fukushima [32,33], Carpentar [22], Hopfield [53,54], etc. Hopfield's illuminating contributions have extended the applicability of neuromorphic techniques to the solution of combinatorially complex optimization problems [55]. In the areas of VLSI and opto-electronic implementations, major achievements have resulted from the efforts of Mead [80], Psaltis and Farhat [91], Hecht-Nielson [48], Sage [96], and others. Grossberg [41], Hartley and Szu [45] have classified neural networks to be any system that satisfies (a) nonlinearity, (b) nonlocal, i.e., exhibit long-range interactions across a network of locations, (c) nonstationary, i.e., interactions are reverabative or iterative, (d) has an "energy-like" function that is non-convex. Its functional components include,

(i) *Network architecture or the topology of interconnections* : Though the bulk of neural network models share a common characteristic in that they are defined as an aggregation of densely interconnected individual processing elements or neurons, it has been shown that the topology of the network has significant influence on network capacity [79], functionality [66], stability [24] and convergence. Topologies may range from crossbar switch (e.g., Kohonen [60], Hopfield [53], etc.) layered networks (e.g., feedforward networks [94], laterally inhibited networks [39]), counterpropagation networks [49], recurrent networks or topographic maps (e.g., Pineda [88], Carpentar and Grossberg [22], Barhen and Gulati [12,13] etc.) or topological maps (e.g., Kohonen et al. [61]). Also, of major interest is the direction of propagation of information, i.e., feedforward nets, feedback nets, counterpropagation nets etc., and the nature of interconnections, i.e., inhibitory or excitatory. Detailed implications on various interconnection topologies are discussed in [95].

(ii) *Problem specific neuron definition* : Although a distinguishing feature of neurocomputation is the ability to accomplish tasks through collective processing, the dynamical behavior of such neural systems depends, to a large extent on the functional laws that describe the basic transfer properties or the temporal behavior of each individual neuron. This is also known as as the *activation dynamics* or *neurodynamics*. Most existing

7

neurodynamical formulations are mathematical idealizations to variants of neurobiological models. They can be essentially classified into two categories, *additive-type* (Hopfield [53,54]) or *shunting-type* (Grossberg [38,40]) interactions. In the former class, the firing of each neuron is preceded by a "gated" summation of the activites of all neurons to which it is connected and a thresholding operation to determine if the energy input exceeds the level above which the cell will fire. On the other hand, shunting-type neurons exert a highly nonlinear influence depending on the spatial-interrelationships , on the activations of neurons to which they are connected. For example, a member of this class, e.g., on-center off-surround neurons [22], strength their own excitation while inhibiting all laterally connected neurons.

iv) *Learning rule :* Learning is the process of adaptively evolving the internal parameters e.g., connection weights, in response to stimuli being presented at the input and possibly the output buffer. learning in neural networks may be supervised, e.g., when the desired response is from a knowledgeable teacher the retrieval involves one or more of a set of stimuli patterns that have been repeatedly shown to the system during the training phase. The networks observe the presented inputs, detect the statistical regularities embedded within it and learn to exploit these regularities to draw conclusions when presented with a portion or a distorted version of the original pattern. When a portion of the original pattern is used as a retrieval cue, the task is denoted as *auto-associative* [17,62]. When the desired input is different from the input the learning is referred to as *hetero-associative*. On the other hand, unsupervised learning proceeds with a knowledgeable teacher. An intermediate kind of learning is reinforcement learning where a teacher just indicates whether the response to an input is good or bad [16], how far and in what direction the current output differs from the desired output, the the network is rewarded or penalized depending on the action it takes in response to each presented stimulus. The network configures itself so as to maximize the reward that it receives. Along with the architecture, learning rules forms the basis of categorizing different neural network models. A detailed taxonomy of different types of learning rules is included in Lipmann [74]. Some of basic types of learning rules are of the form :

Correlational Learning : Learning that postulate parameter changes on the basis of local or global information available to a single neuron. Hebbian learning rule is a good example of correlational learning rule, wherein the connection weights are adjusted according to a correlation between the states of the two interconnected neurons. For example, if two neurons were both active during some successful behavior , the connection would be strengthened to express the positive correlation between them.

Error-corrected Learning : These rules work by comparing the response to a given input pattern with the desired response and then modifying the weights in the direction of the decreasing error, e.g., perceptron learning rule [93], Widrow-Hoff [104], back-propagation [1,87,94,95].

Reinforcement Learning Rule : This model does not require a measure of the desired responses, either at the level of a single neuron or at the level of a network. Only a measure of the adequacy of the emitted response suffices. This reinforcement measure is used to guide random search process to maximize reward, e.g. Barto et al [16].

Stochastic Learning Rule : whereby a network of neurons influencing each other through stochastic relaxation, e.g., Boltzmann Machine [1].

Though the above characterize a neural system, their computational power is distinguished on the basis of the following

[1] *Discrete or Continuous states :* Nature of states of individual neurons is an important axis for classifying neural network models. The number of states can be finite, infinite but countable, or uncountable - forming a continuum [45]. It has been shown that networks with a finite number of states is computationally equivalent to a finite state machine (FSM) if the number of neurons is finite or to a Turing machine (TM) if the number is infinite. If the neuron has a continuum of stable states then it is equivalent to a TM.

[2] *Discrete or Continuous updating :* The nature of time variable in neural computation could be either discrete, i.e., dynamics is approximated by difference approximations to differential equations, or continuous. It has been shown that continuous time networks can resolve temporal behavior [45], which is transparent in networks operating in discrete time. Except for this latter difference the two types are computationally equivalent.

[3] *Fine-grained parallelism* : Neuronal information processing, typically involves combining input information from 10000 input neurons or more, generation of an output signal pulse and its propagation to all axonally connected neurons. It has been also been determined that in general, such pulses are generated over intervals of the order of few milliseconds, and need to be averaged over tens of milliseconds to determine the average rate of firing on the axon. The question that is of interest is as to how the brain can respond to complex stimuli in fractions of a second while individual neural integration requires times of the order of 100 ms ? Given such a scenario, it is reasonable to assume that the relative sluggishness of individual neurons is overcome by the use of massive parallelism in the nervous system. This conviction has now, not only become entrenched as an essential desirable of all neural systems, but also responsible for a lot of excitement that surrounds this discipline. Neural networks are rapidly becoming an attractive paradigm for formulating algorithms, amenable to fine-grain computer architectures.

[4] *Dynamics: Synchronous or Asynchronous* : Another property which significantly affects network performance is the degree of correlation between times of interrogation of individual neurons or the neurodynamics regimes. The updating regimes of existing models may be classified into two basic algorithmic modes [54]. A *concurrent synchronous* mode, where all neurons are updated simultaneously and an *asynchronous* mode, wherein only one randomly selected neuron is allowed to update its state on the basis of its inputs. The firing decision for a particular neuron is allowed only after state information has been received from all other neurons to which it is connected. However, computational connotation of asynchronicity implies uncoordinated systemwide concurrent activity, while the biological manifestation of global asynchrony results from delays in nerve signal propagation, refractory periods and adaptive thresholding [72,73]. So, under such a framework, neural network models do not compute using a systemwide on-off uncoordinated switching (with random delays) of individual neurons. In fact, asynchrony as discussed by [53], in the context of existing artificial neural networks is essentially a "sequential randomness", lacking implicit concurrency. To overcome this limitation, Barhen and Gulati [11] introduced a mathematical framework based on contracting operators, for reconditioning artificial neural network algorithms such that their embodiments are concurrently asynchronous, i.e., arbitrary sets of neurons can be updated at any time.

[5] *Trainability* : Their paradigmatic strength for potential applications, which require solving intractable computational problems or adaptive modeling, arises from their spontaneous emergent ability to achieve *functional synthesis*, and thereby learn non-linear mappings [13,43], and abstract spatial [22,39], functional [66] or temporal [65] invariances of these mappings. Thus, relationships between multiple continuous-valued inputs and outputs can be established, based on a presentation of a large number of representative examples. Once the underlying invariances have been learned and encoded in the topology and strengths of the synaptic interconnections [13,111], the neural network can generalize to solve arbitrary problem instances. Since the topological mappings for problem-solving are acquired from real-world examples, network functionality is not limited by assumptions regarding parametric or environmental uncertainty, that invariably limit model-based adaptive computational strategies [13]. Insightful results regarding the number of training samples required to perform a function successfully, size of the network needed and training time are included in [43].

[6] *Distributed Information processing*

In summary, the computation is emergent, i.e., computationally useful properties such as generalization, classification, association, error correction, time sequence retention etc, emerge as collective properties of such neural systems with large number of simple units. When viewed individually, the dynamics of each neuron bears little semblance to task being performed.

3.   Neural Networks for AI Modeling

Over the past three decades, researchers have charted the ground in the areas of pattern recognition, adaptive learning, perception, sensory-motor control; yielding a good assessment of what is hard and what is easy. Although both have similar goals, there is not much overlap between the projected capabilities of neuromorphic systems and those of computer science/AI [48]. The basis of both paradigms could be traced back to hypotheses of Weiner [103], Leibniz [67] and. They identified human beings as goal-seeking, complicated machines composed of an "intelligent" brain and motor systems. It is able to detect errors, change course, and adapt its behavior so that achievements of goals is more efficient. Subsequent development of intelligent systems has pursued two distinct schools of

11

thought; *AI or symbolic* and *neurobiological or sub-symbolic*. AI researchers concentrated on what the brain did irrespective of how it was accomplished biologically, while the latter focussed on how the brain performed. As an heir to the rationalist reductionist tradition in philosophy [27,84,107,109], AI assumes that there is a fundamental underlying formal representation, and logic that mirror all primitive objects, actions and relations that make up the world, and that intelligent behavior is explicable in scientific terms. Its proponents hypothesize that once such a representation were to become available, the operations of human cybernetic machinery could be fully described in terms of mathematical theorems and formal logic. All knowledge could be formulated into rules, and behavioral aspects of human reasoning and perception could be emulated by following rules or manipulating symbols, without regard to varying interpretations of symbols. Further, intelligent behavior arises from combining symbols in patterns that were not anticipated when the rules were written. Expert systems are product of such a line of investigation. However, over the years AI researchers have unsuccessfully struggled against fundamental issues categorized as the Coding Problem, Category Problem, Procedure Problem, Homunculus Problem, Developmental Problem [92], etc., suggesting recourse to alternate scientific paradigms.

On the other hand, neural network community argues that logical reasoning is not the foundation on which cognition is based, but instead, an emergent behavior that results from observing a sufficient number of regularities in the world. Its theoretical underpinnings lie in biological detail and rigorous mathematics, in an attempt to discover and validate principles that make intelligence possible, by observing existing intelligent systems, i.e., the brain. They hold the view that cybernetic machinery is built from many simple nonlinear interacting elements - neural networks that store knowledge in their internal states and self-organize in response to their environments. Intelligent behavior, then manifests from collective interactions of these units.

While AI or the Symbolic community also treated human brain as an hierarchical system of components that obey laws of physics and chemistry, and could be described as solutions to mathematical equations relating computable functions over the inputs and outputs of neurons, it assumed that given a sufficient amount of information, i.e., computing power, neuronal dynamics, one could compute a person's next state. However, it ignored the framework of interpretation, "context-sensitivity", within which the humans process information, make commitments and assume responsibility. Instead, its primary focus became to design rule-systems that processed symbols without regard to their meanings. Thus, it completely ignored the considerable amount of subsymbolic or subconscious processing precedes our conscious decision making, and subsequently leads to the filtration out of infinity of situations so that the appropriate rule may be used. In sharp contrast, rather than creating logical problem-solving procedures, neural network researchers use only an informal understanding of the desired behavior to construct computational architectures that can address the problem, thereby eliminating the fundamental AI limitation, i.e., context sensitivity [92]. Unlike AI, where there is no recognition, recall and reminding, neural networks focus on association, units and patterns of activation.

Thus artificial neural networks attempt the development and application of human-made systems that can emulate the neuronal information processing operations, e.g., real-time high performance pattern-recognition, knowledge-processing for inexact knowledge domains, and precise sensory-motor control of robotic effectors, that computers and AI machines are not suited for. They are highly amenable for tasks where a holistic overview is required [109], i.e., abstract relatively small amounts of significant information from large data streams as in speech recognition or language identification. On the other hand digital computers and AI are ideal for algorithmic, symbolic, logical and high precision numeric operations that neural networks are not suited for [48]. So in summary, these fields complement each other, in that they approach the same problems, but from different perspectives. Some of the areas to which neural networks have been successfully employed include :

Adaptive Pattern Recognition and classification: for tasks involving invariant perception [63] with respect to stimulus equivalence, position, shift, rotation, perspective, partial occlusion, marring etc. Applications are range from remote sensing, medical imaging, target tracking and classification, computer vision to rudimentary artificial perception organs such as cochlea [80], retina, and modeling of the olfactory bulb.

Approximation of mathematical mappings: In recent years plethora of learning algorithms, e.g., back-propagation, drive-reinforcement [16] learning, singularity interaction dynamics [12,13] etc., have been proposed that can abstract spatial, functional, statistical or temporal invariance from within a set of training samples. Such capability has been applied towards time-series prediction [65], target tracking, sensory-motor control, e.g., neurokinematics and neurodynamics of robot manipulators, adaptive control [4,12], etc.

13

Estimation of probability distributions : has been successfully applied to adaptive signal processing applications involving uncertainty propagation and bandwidth reduction, e.g., speech compression and language identification.

Combinatorial optimization : Following Hopfield and Tank's contribution, several researchers have applied neural networks to obtaining "good", approximate solutions to traditionally hard, multiextremal problems in combinatorics including, multiprocessor scheduling and load balancing [14,15], resource scheduling [44], graph-theoretic problems and path planning.

Adaptive Knowledge Processing Among the major expectations of neural networks is their potential ability to reconstruct or infer from partially defined relational structures for the searching of relevant structures from memory, e.g., Feldman's Connectionist learning architecture [29,30,106]. This then implements a "massively parallel connectionist expert system.

5. Case Study : Neuromorphic Combinatorial Optimization

Combinatorial optimization typically involves finding the minimum or maximum values of an objective function of very many independent discrete variables subject to nontrivial constraints. The objective function represents a quantitative measure of "goodness" of some complex system and depends upon the detailed configuration of the system. Examples include VLSI routing and module placement problem, optimal query decomposition for distributed databases, scheduling problems,

traveling salesman problem. etc. All *exact* methods for determining an optimal to the above problem are known to be NP-Complete, i.e., their computational complexity increases exponentially with the number of variables in the system. Conventionally heuristic methods, e.g., branch and bound, divide and conquer, iterative improvement, etc. have been applied to obtain "near optimal" solutions, with computational requirements

proportional to small powers of N. However, these techniques fail to provide a satisfactory real-time throughput for large-scale application problems with complex competing constraints.Hence, the motivation to exploit massive concurrency, collective computational ability and robustness offered by neural networks to solve difficult problems in combinatorial optimization.

The key steps in attempting to solve an optimization problem with neural networks are :

[1] Construct a generalized objective function, encompassing all network, problem-specific and environmental constraints, such that its global minimum corresponds to the optimal legal solution.

[2] Choose an appropriate representation for encoding the problem in terms of neural variables.

[3] "Translate" the objective function in terms of the above representation.

[4] Construct the solution algorithms, based on either analytical forms for the network synaptic efficacies [55], discrete perturbation expressions [15], simulated annealing [14,44,99], etc.

[5] Provide mechanisms to escape from local minima.

In the sequel we present an efficient combinatorial optimization methodology for obtaining approximate solutions, to the NP-Complete, uniprocessor deterministic scheduling problem in real-time. Given a set of N_T tasks alongwith their computational requirements, arbitrary precedence-constraints among them and their completion deadlines, a neuromorphic model is used to construct a non-preemptive processing schedule, such that their total waiting time and the penalty due to tardiness are minimized. A compact neuromorphic data structure is used to reduce the computational complexity, in terms of number of neurons from $O(N_T N_P)$ to $O(N_T)$. We also discuss an implementation, both, in terms of algorithms based on fast simulated annealing and continuous synchronous neural networks in conjunction with the method of Lagrangian multipliers [89,101]. To illustrate the computational efficacy of our methodology we presents our results for scheduling the highly unstructured inverse dynamics problem of a serial link robot manipulator.

5.1 Illustrative Example : Deterministic Scheduling

Here we present a methodology for addressing the Deterministic Scheduling under hard deadlines in single server and multiserver systems in the light of above objectives. In general, this problem can be defined as follows: Given a set of n heterogeneous tasks, denoted by $J_1, J_2, \cdots, J_n$ to be executed on a single-server system, the problem is to determine an execution sequence which minimizes the waiting time and the tardiness in task-completion. Since the schedule is deterministic all information, required to express the characteristics of the problem is available *a priori*. Each task is associated with a completion deadline, denoted by d_i, by which it must be completed or else a penalty is incurred for the tardiness in completion. To account for the overshoot in completion deadline, the cost term for each task is augmented by an amount denoted by L_i , where

$$L_i = \begin{cases} TB_i + TX_i - d_i & \text{if } TB_i + TX_i > d_i \\ 0 & \text{otherwise} \end{cases} \qquad (5.1.1)$$

where TB_i denotes the time at which execution begins for task i, and TX_i refers to its computational requirement.

Another requirement is the minimization of residence time or the total waiting time spent in the server queue. This is achieved by favoring the execution of shorter jobs over longer ones and reducing the idling cycles. In addition, there are precedence-constraints among the tasks which can be expressed as follows; if the task subset $J_k, \, .. \, , J_o$ denoted by T_i is a precedence constraint of task set, T_j with $J_p, \, .. \, , J_u$ then no task in T_j can start executing until all tasks in T_i have completed. In addition, there are certain physical constraints inherent in the single server scheduling problem which need to be preserved. For example, task overlap is not permitted as there may be utmost one task being served at any time instant. In the general case, e.g. during scheduling tasks on a tool machine at the shopfloor, different schedules may also have different setup costs depending upon the degree of task homogeneity and the type of tool setup required. In such cases the task setup time needs to be taken into account.

Though simple to describe, the solution to the above single server scheduling problem is computationally very hard. It belongs to the NP-complete class of problems [23,34] and its complexity grows exponentially as the number of jobs is increased, arbitrary precedence constraints are present, or, if task preemption is allowed. The complexity of this problem has been analyzed in detail by Lenstra et al [68]. In the past, a variety of elegant deterministic scheduling algorithms have been presented which employ graph-theoretic techniques, quadratic programming, network-flow analysis, branch-and-bound methods, etc. [36,37,68]. Our focus is however on the use of neuromorphic techniques for solving this problem in a general context, i.e., handling task sets with a variety of non-trivial constraints such as completion deadlines, task priorities and precedence constraints. The use of neuromorphic techniques is highly attractive in that they generate conveniently and efficiently, acceptable solutions in real-time.

5.2.   Neuromorphic Representation

The solution to the non-preemptive deterministic scheduling on uniprocessor systems involves determining the sequence in which the tasks are to be executed. The strategy presented by Hopfield and Tank [55], uses the digital output states of N_P bistate neurons to encode the relative position of each of the N_T tasks in the schedule. So for a schedule involving N_T tasks requiring a total of N_P time units for completion, $O(N_T * N_P)$ neurons would be needed if their [55] binary permutation matrix were to be used. However, non-linear neural networks for large task sets derived from that representation and involving several degrees of freedom have been found to converge at unacceptable rates [15]. The major reason for slow convergence being that their [55] representation introduces additional contributing terms in the neuromorphic objective function leading to an increase in the number of degrees of freedom for the system and its computational complexity. For example, the binary permutation matrix based representation requires that there be utmost one 1 in each row and each column. This corresponds to an implicit requirement, i.e., there may be utmost one task executing on the server at any time.

Alternate representations suggested by Fox and Furmanski [31] and Orland [86], based on the binary expansion of encoded neural variable, are used to reduce the computational complexity and constraint space. Instead of specifying relative task positions, an absolute time scale is used to denote the starting time for individual tasks. For example, the starting time of the $n-$tho task is denoted by

$$TB_n = \sum_{i=1}^{K} (1 + V_{ni}) 2^{i-1} \tag{5.2.1}$$

where $K = log_2 N_P$. For a schedule with N_T tasks that requires a total time of N_P units for completion, $log_2 N_P$ neurons are used to encode the starting time for each task. Using such a scheme requires $O(N_T \, log N_P)$ neurons. However, if graded-response state neurons [54] are chosen, then the starting time for each neuron could be encoded to correspond to the stable state of a neuron, i.e., at convergence to a stable attractor each neuron's output will correspond to the starting time of a task and it will take a value between 0 and N_P. Under this schema, one may choose to normalize the neuron's output to generate a continuous output in the interval [0,1].

The above representation results in a more compact system of equations needing O(N_T) neurons. In addition, the system has a fewer number of degrees of freedom than the previous one and thus is more amenable to rapid convergence and constraint satisfiability as it obviates the need for representation constraints.

5.3 Scheduling Energy Function

The algorithm for solving the single server scheduling problem involves minimizing a neuromorphic energy function quantifying the feasibility and optimality constraints in terms of the neural representation over a combinatorial state space corresponding to all possible task schedules. The energy terms contributing to the objective function are formalized below:

With each task is associated a completion deadline by which it must be completed or else a *tardiness* penalty is incurred. The motivation here is to derive a schedule whereby all tasks are completed prior to their specified deadlines, provided such a schedule is feasible. The contribution to the objective function, E_T, due to tardiness in completion of task i is given by bigskip

$$E_t = W_t \sum_i L_i{}^2 H(L_i) \tag{5.3.1}$$

where

$$L_i = TB_i + TXi - D_i \tag{5.3.2}$$

and $H(L_i)$ is the Heaviside function with value equal to one if $L_i > 0$ and zero otherwise. It may be noted that the canonical form of the neural energy function in the Hopfield Model, requires that all energy terms be differential, quadratic functions, in terms of the neural variables. However the presence of Heaviside function, which is discontinuous at zero, in the tardiness energy renders it to be nonquadratic in nature and forces us to seek alternate methods for minimization of neuromorphic energy function.

The energy term due to waiting time is given by

$$E_w = w \sum_i [\frac{(TBi * PR_i)}{TX_i}]^2 \tag{5.3.3}$$

In accordance with our representation TB_i, the output of neuron i and PR_i denotes the priority of task i. This function induces reduction of total queue waiting time for the tasks by favoring shorter and high priority tasks over lengthy and low priority tasks.

We further assume that the execution of tasks is bounded by precedence constraints, such that for each task, i, there is some set of tasks π_i, possibly empty, all of which must be completed before the execution of task i can be initiated. The contribution to the neuromorphic energy due to this constraint is included via the expression,

$$E_p = W_p \sum_i \sum_{j \in \pi_i} \Phi_{ij}^2 H(\Phi(ij)) \qquad (5.3.4)$$

where

$$\Phi_{ij} = TB_j + TXj - TB_i \qquad (5.3.5)$$

The function Φ denotes the unsatisfiability in precedence constraint due to an overlap between a task and its predecessor in the generated schedule or the initiation of a task before all its predecessors are completed. The minimization of waiting time and precedence constraint energies tend to cluster tasks together, introducing task overlap. However, in single server systems there can be atmost one task being processed by the server at any instant. Since this is a physical constraint we provide a crunching routine which eliminates this overlapping among tasks and also removes server idle time when the schedule determination algorithm described below, has converged.

The solution to the task scheduling problem then corresponds to the minimization of the energy functions $E_t + E_w$, alongwith the disappearance of the contribution due to the unsatisfiability of precedence constraints given by E_p. Thus, this reduces to a constrained optimization problem, and may be solved using the method of Lagrangian multipliers. The total energy to be minimized is given by,

$$E = E_t + E_w + \lambda Ep \qquad (5.3.6)$$

where λ is the Lagrange multiplier.

5.4 Minimization of Neuromorphic Energy

Minimization of the objective function, E, derived in Eqn (5.3.6), can be achieved using a variety of techniques, including continuous synchronous and asynchronous neural networks [55], discrete asynchronous neural networks [15], nonlinear differential equations, projected gradient method [89] and classical or fast simulated annealing algorithms [99]. Here we discuss an approach based on synchronous neural network algorithm for the minimization of the scheduling energy function.

The algorithm is based on the projected gradient method in conjunction with the Lagrangian multiplier used for constrained optimization problems. In this method the state of the neuron varies between 0 and N_P inclusive, and all neurons update simultaneously. Essentially this is an iterative procedure on the lines of Projected Gradient Method, which is usually used for minimization of quadratic functions where the variables can be bounded by an N-dimensional cube. A starting point is arbitrarily chosen in the interior of the cube and iteration steps are carried out in the direction of steepest descent with respect to the energy function, E_tot. If a step is not possible in the direction of the steepest descent, i.e there is no additional reduction in energy as a consequence of changing the output of any neuron then it is projected onto the boundary of the feasible region and iterations are continued in the direction of the projection.

In order to ensure that the network converges to a stationary point which corresponds to the local minima of E_{tot}, Lyapunov's stability criteria requires the energy function, i.e., Eqs. (5.3.6) to be monotonically decreasing in time. Since in our model the internal dynamical parameters of interest are the starting times, TB_n and the Lagrange multiplier λ, this implies that

$$\frac{\partial E_{tot}}{\partial t} = \sum_i \left(\frac{\partial E_{tot}}{\partial TB_i} \cdot \frac{\partial TB_i}{\partial t} \right) + \left(\frac{\partial E_{tot}}{\partial \lambda} \cdot \frac{\partial \lambda}{\partial t} \right) \qquad (5.4.1)$$

One can always choose

$$\dot{TB_i} = -\tau \frac{\partial E_{tot}}{\partial TB_i} \qquad (5.4.2)$$

where τ is an arbitrary but positive time-scale parameter. Then substituting in Eqn. (5.4.1) we have

$$\frac{\partial E_{tot}}{\partial t} = -\tau \sum_i \left(\frac{\partial E_{tot}}{\partial TB_i} \right)^2 + \left(\frac{\partial E_{tot}}{\partial \lambda} \dot{\lambda} \right) \qquad (5.4.3)$$

Using Eqn. (5.3.6) the dynamic equations according to which the iterations are carried out are given by

$$\frac{\partial TB_i}{\partial t} = - \frac{\partial E_{tot}}{\partial TB_i} = - \frac{\partial (E_t + E_w)}{\partial TB_i} - \lambda \frac{\partial E_p}{\partial TB_i}. \qquad (5.4.4)$$

Adopting the heuristic recommended by Platt and Barr [], we have

$$\frac{\partial \lambda}{\partial t} = + \frac{\partial E_{tot}}{\partial \lambda} = \pm E_p \qquad (5.4.5)$$

Using the definitions Eqs. (5.3.1), (5.3.3) and (5.3.4) the component terms $\frac{\partial E_t}{\partial TB_i}$, $\frac{\partial E_w}{\partial TB_i}$ and $\frac{\partial E_p}{\partial TB_i}$ can be computed as

$$\frac{\partial E_t}{\partial TB_i} = 2W_t \, L_i H(L_i) \qquad (5.4.6)$$

$$\frac{\partial E_w}{\partial TB_i} = 2W \left(\frac{TB_i \cdot PR_i}{TX_i} \right) \left(\frac{PR_i}{TX_i} \right) \qquad (5.4.7)$$

and

$$\frac{\partial E_p}{\partial TB_i} = 2Wp \left\{ - \sum_{j \in \pi_i} \Phi_{ij} \, H(\Phi ij) + \sum_{j \in \bar{\pi}_i} \Phi_{ji} \, H(\Phi ji) \right\} \qquad (5.4.8)$$

where $\bar{\pi}_i$ denotes the set of tasks reachable from i.

Thus, $\partial E_{tot}/\partial t$ is negative at each iteration, one can expect the system to converge, resulting in valid schedule. This state corresponds to one of the local minima of E_{tot} in the n-dimensional space. However in our case, since E_p and E_t are not continuous as a consequence of the presence of the Heaviside function, the eqn (5.4.7) cannot indicate the direction of changes in the energy function. In all cases derived above, we obtained an energy decrease, although $\partial E_t ot / \partial t$ was positive in some iterations. If this does occur and the energy increases in some iterations, it would not change the general trend of the energy to decline. Further this occasional increase of energy is instrumental in leading to valid solution in that it forces the state trajectory to deviate and escape from one basin of attraction, which may not lead to constraint satisfaction, to another one which does. This hill climbing as a consequence of occasional increases in the total energy causes changes in the basins of attraction [59].

As far as the Lagrange multiplier is concerned, in cases like ours, where the constraint function E_p is always positive, one may use either sign in eqn (4.2.2). However, one should note that in each case the value of λ will converge to a constant steady state value and:

a. In the case of a minus sign, the initial value of λ should be high enough such that the value of λ does not become negative, a because negative value of λ will cause the solution to diverge. This case is analogous to providing a high weight for the constraint function at the initiation of the iterations in order to force the solution to satisfy the constraints. The weight monotonically decreases along with the iterations and allows other components of the energy function to be minimized.

b. On the other hand if the positive sign is selected then the initial value of λ can be chosen randomly in $[-\infty$, $+\infty]$. In this case the negative value of λ would not cause any problems but it will take longer to satisfy the constraints. Thus, it is not recommended to initialize λ with a negative value. If the iterations are started with low weights on the constraint functions then the network converges to a good solution. By increasing the value of λ at each iteration, depending on the value of the unsatisfied constraint function , one can force the state trajectory to towards valid solutions.

5.5. Simulations

We now address some of the practical considerations related to the simulation of neuromorphic scheduling algorithms. In order to provide a context and explore the efficacy of our methodology in scheduling large task sequences with non-trivial constraints, we analyzed the problem of scheduling the decomposed solution (in the Newton Euler formalism) to inverse dynamics equations of the Stanford 6-DOF robot manipulator. This problem was chosen as a benchmark because it is relatively simple, while exhibiting complex precedence relations and structural irregularity characteristics of interest. Several state-of-the-art formalisms are currently available to efficiently solve the inverse dynamics problem of a serial link manipulator. The dynamic behavior of a manipulator is modeled by a set of coupled, highly nonlinear equations of motion and the aim is to calculate the forces or torques based on known motion. Detailed derivations and task decompositions for the above system with 88 tasks can be found in Barhen et al [15], while intertask dependencies are discussed in [14].

The scheduling system comprised of a neural network with 88 graded-response state neurons, with arbirtrary starting states. For this network, the output of a neuron i corresponded to the starting time for task$i$. A number of optimization methodologies were attempted, e.g., fast simulated annealing (discussed in [44]), synchronous and asynchronous neural network algorithms. Figure 4 shows the optimization results obtained using the implementation based on continuous synchronous neural networks, discussed in the preceding section. This graphs displays the monotonic variation of the total neuromorphic energy, E_{TOT} (5.3.6) as the schedule is generated. Note the rapid convergence to a stable state in 0.4s effective time (discretization time constant chosen, $\Delta = 0.001$). Although, this algorithm is based on gradient-descent, the algorithm is susceptible to convergence to local minima, the schedule arrived at as a result of our synchronous minimization satisfied all the precedence and tardiness constraints. In Gulati et al [42], we include a variant of the synchronous optimization algorithm, derived in strict conformity to Lyapunov stability requirements and based on non-lipschitzian dynamics, that can speed-up the minimization process by upto two orders of magnitude. Figure 5 presents the normalized evolution of waiting time component contribution, E_{WAIT}, tardiness component contribution, E_{TAR} and component contribution due to precedence constraints E_{PIX}. It can be immediately deduced from the graph, that for our existing formalism the satisfiability of precedence and communications constraints completely dominates the overall system performance. When the system converges to a stable state, the schedule can be trivially obtained by reading the neuronal outputs.

6.   Summary

The resurgence of neural networks as a computational discipline, fueled by the cross-fertilization of technology, science and cybernetic intent, is indeed an exciting event. It is part of increasing trend towards massively, parallel, distributed computing and cooperative phenomena. The field has been energized by developments in cognitive science, computational neurobiology, linguistics, dynamical systems, physchophysics, computer science, and especially the advances in VLSI, analog hardware and networking. This is in fact an opportunity to address the unmet expectations of AI and classical cybernetics. But, while the goals are high, so is the challenge, and thus the implicit impatience and temptation to expect imminent success in the short term. It must be remembered that biological intelligence, cognitive and perceptual machinery has evolved over tens of thousands of years,

FIGURE 4

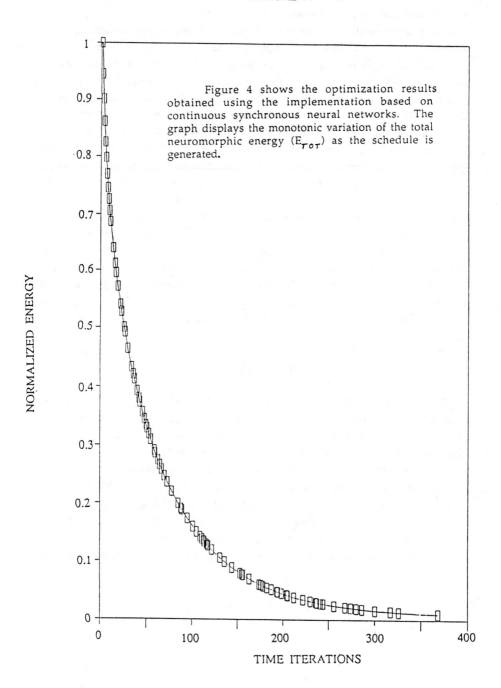

Figure 4 shows the optimization results obtained using the implementation based on continuous synchronous neural networks. The graph displays the monotonic variation of the total neuromorphic energy (E_{TOT}) as the schedule is generated.

FIGURE 5

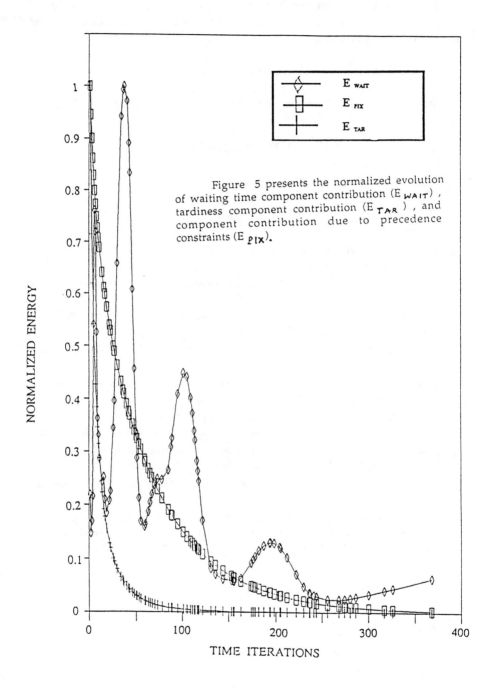

Figure 5 presents the normalized evolution of waiting time component contribution (E_{WAIT}), tardiness component contribution (E_{TAR}), and component contribution due to precedence constraints (E_{PIX}).

and their emulation will not be possible in the near-term neural network and AI research on VLSI silicon chips [92]. However, this discipline with its ability to address fundamental problems such as learning, sensory-motor control, machine vision, analogical reasoning, etc., offers a lot of possibilities.

Acknowledgements

The authors wish to acknowledge useful discussions with N. Toomarian, C. Jorgensen. and S. Gulati.

REFERENCES

[1] Ackley, D.H., Hinton, G.E. and Sejnowski, T.J (1985). "A Learning Algorithm for Boltzmann Machines", *Cognitive Science, 9*, 185.

[2] Adrian, E.D., (1914). "The All-or-None Principle in Nerve", *J. Physiol., Lond.*, XLVII, (6), 460-474.

[3] Aho, A.V., Hopcroft, J.E. and Ullman, J.D. (1974). *Design and Analysis of Algorithms*, Addison-Wesley, Reading, MA.

[4] Albus, J.S. (1971). "A Theory of Cerebellar Function", *Mathematical Bioscience, 10*, 25.

[5] Amari, S.-I. (1972). "Characteristics of Random Nets of Analog Neuron- Like Elements", *IEEE Trans. Sys., Man, and Cybern*, SMC-2 (5), 643-657.

[6] Amari, S-I. (1974). "A Method of Statistical Neurodynamics", *Biological Cybernetics*, 14, 201.

[7] Amari, S-I., and Arbib, M.A. (1982). *Competition and Cooperation in Neural Nets*, Springer-Verlag, New York, NY.

[8] Amit, D.J., Gutfreund, H. and Sompolinsky, H. (1985). "Storing Infinite Numbers of Patterns in a Spin-Glass Model of Neural Networks", *Phys. Rev. Lett. 55*, 1530-1533.

[9] Arbib, M. and Amari, S.I (1985). "Sensory-motor transformations in the brain (with a Critique of the Tensor Theory of the Cerebellum", *J. Theo. Biol.*, 112, 121-155.

[10] Ballard, D.H. (1986). "Cortical Connections and Parallel Processing", *Behav. Brain Sc.*, *9*(1), 67.

[11] Barhen, J. and Gulati, S. (1989). "Chaotic Relaxation Neurodynamics in Concurrently Asynchronous Neural Networks", in *Proc. 1989 Int'l Joint Conf. in Neural Networks*, Washington, D.C., II, 619-626.

[12] Barhen, J. and Gulati, S. (1989). "Self-Organizing Neuromorphic Architecture for MAnipulator Inverse Kinematics", *NATO ASI* (in press).

[13] Barhen, J., Gulati, S., and Zak, M. (1989). "Neural Learning of Constrained Nonlinear Transformations", *IEEE Computer*, 22, 67-76.

[14] Barhen, J. and J.F. Palmer (1986). "Hypercube in Robotics and Machine Intelligence", *Comp. Mech. Engg.* , 4(5), 30.

[15] Barhen, J., Toomarian, N. and Protopopescu, V. (1987). "Optimization of the Computational Load of a Hypercube Supercomputer Onboard a Mobile Robot," *Applied Optics*, 26, 5007-5014.

[16] Barto, A.G. Sutton, R.S., and Anderson, C.W. (1983). "Neuronlike Adaptive Elements that can Solve Difficult Learning Control Problems", *IEEE Trans on System, Man and Cybernetics*, SMC-13, 834-846.

[17] Baum, E.B., Moody, J. and Wilczke, F. (1987). "Internal Representations for Associative Memory", *Inst. for Theor. Phys.*, University of California. Also to appear in *Biological Cybernetics*

[18] Blomfield, S. and Marr D. (1970). "How the Cerebellum May Be Used", *Nature 227*, 1224-1228.

[19] Braham, R., and Hamblen, J.O. (1988). On the behavior of some Associative Neural Networks, *Biol. Cybern*, 60, 145-151.

[20] Caianiello, E.R. (1961). "Outline of a Theory of Thought Processes and Thinking Machines", *J. Theor. Biol. I*, 204-235.

[21] Cajal, S., and Ramon, Y. (1908). *Histology du Système Nerveux*, Madrid, CSIC (reprinted in 1972).

[22] Carpenter, G.A., and Grossberg, S. (1985). "A Massively Parallel Architecture for a Self-Organizing Neural Pattern Recognition Machine", *Computer Vision, Graphics, and Image Processing*, 37, 54-115.

[23] Coffman, E.G. (1976). *Computer and Job Shop Scheduling Theory* , J. Wiley, New York.

[24] Cohen, M.A., and Grossberg, S. (1983). "Absolute Stability of Global Pattern Formation and Parallel Memory Storage by Competitive Neural Networks", *IEEE Trans. System, Man and Cybernetics*, SMC-13, 815-826.

[25] Denker, ed. (1986). "Proceedings of the AIP Conf. Neural Networks for Computing", 151.

[26] Denning, P.J. (1988). "Blindness in Designing Intelligent Systems", *American Scientist*, 76, 118-120.

[27] Dreyfus, S.E. (1987). "Neural Nets: An Alternative Approach to AI", *Applied AI Reporter*, 6-15.

[28] Durbin, R. and Willshaw, D.J. (1987). "An Analogue Approach to the Travelling Salesman Problem Using an Elastic Net Method", *Nature 326,* 689-691.

[29] Feldman, J.A., and Ballard, D.H. (1982). "Connectionist Models and Their Properties", *Cognitive Science*, 6, 205-254.

[30] Feldman, J.A. (1986). "Neural Representation of Conceptual Knowledge ", *Tech. Rep. TR189*, Dept. of Computer Science, Univ. of Rochestor, 1986.

[31] Fox, G. and W. Furmanski. "Load Balancing by a Neural Network", submitted for publication to *Journal of Supercomputing* .

[32] Fukushima, K. (1975). "Cognitron: A Self-Organizing Multilayered Neural Network, *Biological Cybernetics*, 20(3/4), 121-136.

[33] Fukushima, K. (1980). "Neocognitron: A self-Organizing Neural Network Model for a Mechanism of Pattern Recognition unaffected by shift in Position, *Biological Cybernetics*, 36(4), 193-202.

[34] Garey, M.R. and D.S. Johnson (1979). *Computers and Intractability: a Guide to the Theory of NP-Completeness*, Freeman, San Francisco, CA.

[35] Gemen, S. and Gemen, D. (1987). "Stochastic Relaxation, Gibbs Distribution, and the Bayesian Restoration of Images", Readings in Computer Vision, Morgan Kaufmann Publishers, Inc., 564-584.

[36] Gonzalez, J. (1977). "Deterministic Processor Scheduling," *ACM Computing Surveys* , 9(3), 173.

[37] Graham, R.L., Lawler, E.L. Lenstra, J.K., and A. H. G. R. Kan (1979). "Optimization and Approximation in Deterministic Sequencing and Scheduling: A Survey," *Annals Discrete Math.* , 5, 169.

[38] Grossberg, S. (1973). "Contour Enhancement, Short Term Memory, and Constancies in Reverberating Neural Networks", *Studies in Appl. Math.* LII (3), 213-257.

[39] Grossberg, S. (1986). "Some Nonlinear Networks Capable of Learning a Spatial Pattern of Arbitrary Complexity", *Proc. of Nat'l Acad. of Sci.*, 59, 368, 372.

[40] Grossberg, S. (Ed.) (1987). "The Adaptive Brain", I and II, Amsterdam: Elsevier/North Holland.

[41] Grossberg, S. (1988). "Nonlinear Neural Networks: Principles, Mechanisms and Architectures", *Neural Networks*, 1(1), 17-61.

[42] Gulati, S., Barhen, J. and Iyengar, S.S. (1989) "Computational Learning Formalisms for Manipulator Inverse Kinematics", *Proc. of NASA Workshop on Space Telerobotics*, Pasadena, CA.

26

[43] Gulati, S., Barhen, J., and Iyengar, S.S. (1990). "Neurocomputing Formalisms for Learning and Machine Intelligence", to appear in *Advances in Computer*, ed. M.C. Yovits, Academic Press, New York, NY.

[44] Gulati, S., Iyengar, S.S., Toomarian, N., Protopopescu, V. and J. Barhen (1987). "Nonlinear Neural Networks for Deterministic Scheduling", *IEEE First Int. Conf. on Neural Networks*, IV, 745.

[45] Hartley, R., and Szu, H. (1987). "A Comparison of the Computational Power of Neural Network Models", *Proc. IEEE 1987 Int'l Conf. on Neural Networks*, III, 15-22.

[46] Hofstadter, D.R. (1979). *Godel, Escher, Bach*, New York, Basic Books.

[47] Hebb, D.O. (1949). "Organization of Behavior", New York: Wiley.

[48] Hecht-Nielson, R. (1986). "Performance Limits of Optical, Electro-optical and Electronic Artificial Neural System Processors," *Proc. SPIE*, 634, 277.

[49] Hecht-Nielson, R. (1988). "Applications of Counterpropagation Networks", *Neural Networks*, 1(2), 131-139.

[50] Hinton, G. (1984). "Distributed Representations", *Tech. Report, CMU-CS-84-157*, Dept. of Computer Science, CMU, Pittsburgh.

[51] Holland, J.H. (1975). "Adaptation in Natural and Artificial Systems", *University of Michigan Press*, Ann Arbor.

[52] Hodgkin, A.L. and Huxley, A.F. (1952). "A Quantitative Description of Membrane Current and its Application to Conduction and Excitation in Nerve", *J. Physiol., Lond. 117*, 500-544.

[53] Hopfield, J.J. (1982). "Neural Networks and Physical Systems with Emergent Collective Computational Abilities", *Proc. natn. Acad. Sci. USA 79*, 2554-2558.

[54] Hopfield, J.J. (1984). "Neurons with Graded Response have Collective Computational Properties Like Those of Two-State Neurons", *Proc. Nat'l Acad.Sci.*, 81, 3058-3092.

[55] Hopfield, J.J. and Tank, D.W. (1985). "Neural Computation and Constraint Satisfaction Problems and the traveling Salesman", *Biol. Cybern. 55*, 141- 152.

[56] Ito, M., (1984), "The Cerebellum and Neural Control", Raven, New York.

[57] Jackel, L.D., Graf, H.P. and Howard, R.E. (5077). "Electronic Neural Network Chips", *Applied Optics*, 26(23), 5077-5080.

[58] Kanal, L. and Tsao T. (1986). "Artificial Intelligence and Natural Perception", *Proc. of Intelligent Autonomous Systems*, Amsterdam, 60-70.

[59] Keeler, J.D. (1986). " Basins of Attraction of Neural Network Models ", *Proc. of AIP Conference* , 151, Neural Networks for Computing, Snowbird, UT, 259-265.

[60] Kohonen, T. (1977). "Associative Memory: System Theoretic Approach", Springer-Verlag, Berlin.

[61] Kohonen, T. (1982). "Self-Organized Formation of Topologically Correct Feature Maps", *Biol. Cybern, 43,* 59-70.

[62] Kohonen, T. (1984). "Self-Organization and Associative Memory", Berlin: Springer-Verlag.

[63] Kohonen, T. (1988). "State of the Art in Neural Computing", in *Proc. of IEEE First Int'l Conf. on Neural Networks,* I, 79-90.

[64] Ladd, S. (1985). "The Computer and the Brain: Beyond the Fifth Generation", Bantom Books, New York.

[65] Lapedes, A. and Farber, R. (1986). "A Self-Optimizing, Nonsymmetrical Neural Net for Content Addressable Memory and Pattern Recognition", *Physica D 22,* 247.

[66] Le Cun, Y. (1985). "A Learning Scheme for Asymmetric Threshold Networks", *Proc. Congitiva 85,* 599-607.

[67] Leibniz, J. (1951). "Selections", ed. Philip Wiener, New York: Scribner.

[68] Lenstra, J.K., and Rinnoy Kan, A.H.G. (1978). "Complexity of Scheduling Under Precedence Constraints", *Operations research* , 26(1), 22.

[69] Linsker, R. (1986a). "From Basic Network Principles to Neural Architecture: Emergence of Spatial-Opponent Cells", *Proc. natn. Acad. Sci. U.S.A. 83,* 7508-7512.

[70] Linsker, R. (1986b). "From Basic Network Principles to Neural Architecture: Emergence of Orientation Selective Cells", *Proc. natn. Acad. Sci. U.S.A. 83,* 8390-8394.

[71] Linsker, R. (1986c). "From Basic Network Principles to Neural Architecture: Emergence of Orientation Columns", *Proc. natn. Acad. Sci. U.S.A. 83,* 8779-8783.

[72] Little, W.A. (1974). "The Origin of the Alpha Rhythm", Edinburgh, London: Churchill Livingstone.

[73] Little, W.A. and Shaw, G.L. (1975). "A Statistical Theory of Short and Long-Term Memory", *Beh. Biol. 14,* 115-133.

[74] Lippmann, R.P. (1987). "An Introduction to Computing with Neural Nets", *IEEE ASSP Magazine, 4*(2), 4.

[75] Malsburg, Ch. V.D. (1985). "Self-Organization of Orientation Sensitive Cells in the Striate Cortex", *Bunsenges. phys. Chem 89,* 703-710.

[76] Marr, D. (1969). "A Theory of Cerebellar Cortex", *Jour. Physiology of London, 202,* 437.

[77] P. McCorduck (1979). "Machines who Think", W.H. Freeman and Co., New York.

[78] McCulloch, W.S. and Pitts, W.H. (1943). "A Logical Calculus of the Ideas Immanent in Nervous Activity," *Bulletin of Mathematical Biophysics, 5*,115.

[79] McEliece, R.J. Posner, E.C., Rodemich, E.R. and Venkatesh, S.S. (1987). "The Capacity of the Hopfield Associative Memory", *IEEE Trans. Inf. Theory I*, 33-45.

[80] Mead, C. (1989). *Analog VLSI and Neural Systems*, Addison Wesley, Reading, MA.

[81] Minsky, M., and Papert, S. (1969). "Perceptrons: An Introduction to Computational Geometry", Cambridge: MIT Press.

[82] Mjolsness, E. and Sharp, D.H. (1986). "A Preliminary Analysis of Recursively Generated Networks", In *Neural Networks for Computing* (ed. J.S. Denker, E. Mjolsness, D.H. Sharp and B.K. Alpert). *AIP Conf., 151.*

[83] Mjolsness, E. (1987). "Control of Attention of Neural Networks", Yale University.

[84] Newell, A., and Simon, H. (1976). "Computer Science as Empirical Inquiry: Symbols and Search," *Comm. ACM.*

[85] North, G. (1987). "A Celebration of Connectionism", *Nature, 328*, 107.

[86] Orland, H. (1985). "Mean Field Theory for Optimization Problems", *Journal of Physics, Paris*, 46, L673.

[87] Parker, D.B. (1986). "A Comparison of Algorithms for Neuron-like Cells", In: *Neural Networks for Computing* (ed. J.S. Denker). *Proc. AIP Conf 151.*

[88] Pineda, F.J. (1987). "Generalization of Back-Propagation to Recurrent Neural Networks," *Physical Review Letters* , 59, No. 19, 1987, 2229-2232.

[89] Platt, J.C. and Barr, A.H. (1987). "Constrained Differential Optimization", in *Proc. of 1987 IEEE Conf. Neural Information Processing Conference*, Denver, CO.

[90] Poggio, T., Torre, V. and Koch, C. (1985). "Computational Vision and Regularization Theory", *Nature*, 314-319.

[91] Psaltis, D. and Farhat, N.H. (1985). "Optical Information Processing Based on an Associative-Memory Model of Neural Networks with Thresholding and Feedback, *Optical Letters*, 10, 98-100.

[92] Reeke, G.N., and Edelman, G.M. (1989). "Real Brains and Artificial Intelligence", *The AI Debate*, 144-173.

[93] Rosenblatt, F. (1962). "Principles of Neurodynamics, Perceptrons and the Theory of Brain Mechanisms", Spartan Books, Washington, D.C.

[94] Rumelhart, D.E., Hinton, G.E. and Williams R.J., "Learning Representations by Back-Propagating Errors", *Nature, 323*, 533.

[95] Rumelhart, D.E., McClelland, J.L. and the PDP Research Group (1986). "Parallel Distributed Processing", Vol. I and II, MIT Press, Cambridge, MA.

[96] Sage, J.P., Thompson, K. and Withers, R.S. (1986). "An Artificial Neural Network Integrated Circuit Based on NMOS/CCD Principles", *AIP Conf. Proc.*, 151, 381.

[97] Schori, R.M. (1987). "A Case for Neural Networks", *AFOSR Tech. Report.*

[98] Sejnowski, T.J. and Rosenberg, C.R. (1986). "NETtalk: A Parallel Network that Learns to Read Aloud", *Tech. Rep. JHU/EECS-86/01*, Johns Hopkins Univ., Baltimore, EECS.

[99] Szu, H. (1986). "Fast Simulated Annealing", *American Institute of Physics*, Ed. J.S. Denker, New York, NY.

[100] Taylor, W.K. (1956). "Electrical Simulation of Some Nervous System Functional Activities", Information Theory, ed. E.C. Cherry, Butterworths, London.

[101] Toomarian, N. (1988). "A Concurrent Neural Network Algorithm for the Traveling Salesman Problem", Proc. of the Third Conference on Hypercube Concurrrent Computers and Applications, Pasadena, CA.

[102] Waltz, D.L. (1988). "The Prospects for Building Truly Intelligent Machines", *Daedalus*, Boston, Mass.

[103] Weiner, N. (1948). "Cybernetics, or Control and Communications in the Animal and the Machine", John Wiley, New York.

[104] Widrow, B. and Hoff, M.E. (1960). "Adaptive Switching Circuits", *WESCON Convention Record*, 4, 96.

[105] Willshaw, D.J., Buneman, O.P. and Longuet-Higgens, H.C. (1969). " Non-Holographic Associative Memory", *Nature*, 222, 960.

[106] Willshaw, D.J. and Malsburg, Ch. V.D. (1976). "How Patterned Nerual Connections Can be Set Up by Self-Organization", *Proc R. Soc. Lond. B194*, 431-445.

[107] Winograd, T. (1976). "Artificial Intelligence and Language Comprehension", National Institute of Education, Washington, D.C.

[108] Winograd, S. and Cowan, J.D. (1963). "Reliable Computation in the Presence of Noise", MIT Press, Cambridge, MA.

[109] Wittgenstein, L. (1975). "Philosophical Remarks", University of Chicago Press, Chicago, IL.

[110] Zak, M. (1988), "Terminal Attractors for Addressable Memory in Neural Networks", *Physics Letters A*, 133, 18-22.

[111] Zak, M. (1989). "The Least Constraint Principle for Learning in Neurodyamics", *Physics Letters A*, 135, 25-28.

[112] Zak, M. (1989b). "Terminal Attractors in Neural Networks", *Int'l Journal on Neural Networks*, 2(3), 259-274.

2

THE BOLTZMANN MACHINE:
THEORY AND ANALYSIS
OF THE INTERCONNECTION TOPOLOGY

A. De Gloria and S. Ridella

Dipartimento Ingegneria Biofisica Elettronica - Università di Genova

Via All'Opera Pia, 11 A - Genova - Italy

Istituto Circuiti Elettronici - CNR Genova

Via All'Opera Pia, 11 - Genova - Italy

ABSTRACT

The Boltzmann Machine is a very interesting neural network for its similarities with the natural process of statistical thermodinamics, and for its distributed computation abilities, which open wide possibilities of realization both by standard VLSI techniques and by future technologies such as molecular electronics and optics. A review of the theory of the Boltzmann Machine is presented together with an analysis of different interconnection topologies. The representation problem is discussed and it is shown how BM can represent any boolean mapping between any number of inputs and one output. A theoretical result for the parity problem is also shown.

1 INTRODUCTION

Neural networks (NN) are a promising computational model for their capability in modelling and solving problems hardly approachable with traditional techniques such as statistical pattern recognition and artificial intelligence [1].

The main NN attractive is their learning capability by which they can solve problems whose formalization is not known and whose solution is available only in some instances used as training examples. In this sense NN seem to have categorization abilities allowing to extract general rules from particular situations. Several examples exist in literature where NN with few neurons solve in an elegant way simple pattern recognition or signal interpretation problems [1].

Unfortunately NN models require a lot of computing time to be simulated on a sequential machine resulting in a great difficulty to investigate the behaviour of large NN and to verify their ability to solve real problems. So that, the research in the NN field also leads toward their hardware implementation.

Despite the now available impressive large integration ability, VLSI technology imposes constraints on NN hardware implementation.

One of the main point in the debate is about advantages and disadvantages of analogic, digital and hybrid approaches.

The analogic design seems attractive to preserve some pecuraliarities of NN. Moreover some NN components can be easily mapped on silicon.

For example the sigmoid, which is often used as the transfer function of a neuron, can be emulated by using an operational amplifier and synapses are easily modelled by using resistances.

Despite these similarities between NN and the electric circuit theory, the mapping of other functions, mainly related to the learning capability, is still an unsolved problem.

In fact, learning needs memory to store weight values computed with high accuracy in the training stage.

It is well known in the VLSI design that reliable writing memories can be only realized by using a digital approach, since proposed analogic techniques, as the use of gate transistor capacitor, don't assure a long storing of the exact value originally stored in it.

At the present time the precision of the weights seems to not to have easy reproduction by using an analogic approach, too.

Another difficulty in the analogic implementation is that learning requires a supervisor control able to perform decisions about the weight updating: the Finite State Machine approach appears to be the only viable one to implement NN learning control.

For these reasons, at the present time the analogic design of NN is an interesting research field but probably it will not produce in a short time results which allow to experiment large NN, with learning ability.

Digital design is based on well established and reliable techniques and can be used to easily realize cheap NN chips with high performances as compared to software implementation on sequential computer. This approach looks at a NN emulation on silicon without any reference to computational mechanisms and topology.

Digital approach causes constraints on the design of NN, too. Interconnections represent one of the main factor in wasting silicon area and reducing performances, so investigations have to be accomplished to limit the number of synapses in a NN.

Moreover the operations the NN algorithm performs, have to be as simple as possible to avoid the design of complex computing structures. The VLSI technology leads toward distributed control so algorithms requiring a central control must be avoided too.

These considerations lead to choose the Boltzmann machine(BM) as the best candidate for silicon implementation. BM has a distributed control and its neurons can only assume two values 0 and 1, therefore the multiplication of a neuron value by the synapse weight can be implemented with the logical "and" function avoiding the design of a multiplier.

The operation comparing between the sum of synapse weights by the neuron values with the neuron threshold is also avoided and in its place a compare with zero is requested resulting in a simple hardware structure.

The stochastic process involved in the decision about the state of a neuron can be implemented by a look up table avoiding the run-time computing of the exp function.

Despite the advantages of the BM cited above, many implementation problems are still open, mainly regarding the tuning of the topology and learning parameters.

These problems are investigated in this paper through simulation of BM on a digital computer.

2   THE BOLTZMANN MACHINE

In this section a description of the BM algorithm will be given; more details about the mathematical proofs can be found in [1][2[3][12][13][14][15].

A BM is composed of N neurons connected by links and each of them is characterized by a weight (w). Neurons are of digital type and therefore can be only in two states (s): 1 and 0.

Neurons can be subdivided in input (NI), output (NO) and hidden (NH) neurons. About the state of the latter it is not possible to have a-priori informations both in the training and in the testing phase.

The input neurons state is always determined while the output neurons one is known only in the testing phase. The need to have hidden neurons, to allow the representation of any input-output relation, is well-known [4].

An index is associated to each neuron and w_{ij} is the weight of the link connecting the neurons i and j.

The weights are simmetric, such as $w_{ij} = w_{ji}$.

A bias is associated to each neuron and is identified by w_{ii} which is equivalent to a weight between the neuron i and a neuron forced in the state 1.

The process, throught which the state of neurons are determined when an input is applied, is inspired to the thermal annealing of statistical mechanics.

An energy function is defined:

$$E = -1/2 \sum_{i}^{N} \sum_{j \neq i}^{N} W_{ij} S_i S_j - \sum_{i}^{N} W_{ii} S_i \tag{1}$$

and the unknown states are determined by looking for the global minimum of E.

Since this problem is of combinatorial optimization type, continous methods based on gradient search cannot be applied. Therefore well-known simulated annealing techniques [5] must be used.

The simulated annealing is accomplished through traditional Metropolis[10] algorithm. The procedure is based on a random choice of a neuron k and on the computation of the quantity:

$$\Delta E_k = \sum_{j} w_{ij} a_{ij} + W_k \tag{2}$$

that is the variation of energy when the neuron k changes from state (1) to state (0), whose energies are E_1 and E_0 respectively. Since, the probability that neuron k is in state (1) in Boltzmann mechanics is:

$$p_1 = \frac{e^{-E_1/T}}{e^{-E_0/T} + e^{-E_0/T}} = \frac{1}{1 + e^{-\Delta E_k/T}} \tag{3}$$

while the probability of the state (0) is:

$$p_0 = 1 - p_1 \tag{4}$$

T is the value of the temperature considered. The state of neuron k is fixed in the state (0) or (1) according with the computed probabilities. Then another neuron is randomly choosen and the procedure is repeated. The process terminates when the network reaches the thermodynamics equilibrium.

It must be stressed that this technique searches the minimum by using quantities which are local in respect to the choosen neuron. Therefore this technique well match distributed computation requirement.

The strong analogy of the procedure with physical processes gives a hope for a future hardware implementation with very innovative technology, e.g. molecular electronics.

The final problem is to obtain an algorithm to determine the weight and bias values given a known training set of patterns.

The traditional cost function, based on the sum of the squares of the differences between the target and the corresponding output, is not a continuous function of the weights and its minimization becomes a combinatorial problem which can be solved by using the previous described simulated annealing algorithm. This is a global optimization technique which is very expensive for what concerns computation. Moreover, for this particular application, it is not known whether it can be used only through local decisions implying a weight updating. This one is based only on the state of the neurons connected to the considered weight.

A cost function has been defined [2,3] based on an information theoretic measure of the disagreement between the network's internal model and the environment:

$$C = \sum_{\alpha} p(V_\alpha) ln \frac{p(V_\alpha)}{p'(V_\alpha)} \tag{5}$$

where $p(V_\alpha)$ is the probability of the αth state of the input and output neurons when their states are clamped, and $p'(V_\alpha)$ is the corresponding probability when the output neurons are unclamped.

It can be demonstrated [11] that C is always greater that zero except when $p(V_\alpha) = p'(V_\alpha)$. This is a continous function whose minimum can be found by a gradient descent technique:

$$\Delta W_{ij} = -\eta \frac{\partial C}{\partial W_{ij}} \tag{6}$$

Derivation of the cost gradient [3] needs only the computation of $p'(V_\alpha)$ since $p(V_\alpha)$ does not depend on w_{ij}. When the BM is in the phase corresponding to $p'(V_\alpha)$ the input units are clamped and the output units and the hidden units free-run. At equilibrium, the probability distibution over the visible units is given by:

$$p'(V_\alpha) = \sum_{\beta} p'(V_\alpha \bigwedge H_\beta) = \sum_{\beta} e^{-E_{\alpha\beta}/T} / \sum_{\lambda\mu} e^{-E_{\lambda\mu}/T} \tag{7}$$

where V_α is a vector of the states of the visible units, H_β is a vector of the states of the hidden units, and $E_{\alpha\beta}$ is the energy of the system in state $V_\alpha \bigwedge H_\beta$:

$$E_{\alpha\beta} = -1/2 \sum_{i}^{N} \sum_{j \neq i}^{N} W_{ij} S_i^{\alpha\beta} S_j^{\alpha\beta} - \sum_{i}^{N} W_{ii} S_i^{\alpha\beta} \tag{8}$$

The derivation of the energy $E_{\alpha\beta}$ with respect to the weight w_{ij} gives :

$$\frac{\partial e^{-E_{\alpha\beta}/T}}{\partial w_{ij}} = \frac{1}{T} s_i^{\alpha\beta} s_j^{\alpha\beta} e^{-E_{\alpha\beta}/T} \tag{9}$$

The derivative $p'(V_\alpha)$ with respect to w_{ij} is given by:

$$\frac{\partial p'(V_\alpha)}{\partial w_{ij}} = \frac{1}{T} \frac{\sum_\beta e^{-E_{\alpha\beta}/T} s_i^{\alpha\beta} s_j^{\alpha\beta}}{\sum_{\alpha\beta} e^{-E_{\alpha\beta}/T}} - \frac{1}{T} \frac{\sum_\beta e^{-E_{\alpha\beta}/T} \sum_{\lambda\mu} e^{-E_{\lambda\mu}/T} s_i^{\lambda\mu} s_j^{\lambda\mu}}{(\sum_{\lambda\mu} e^{-E_{\lambda\mu}/T})^2} =$$

$$= \frac{1}{T} \sum_\beta p'(V_\alpha \bigwedge H_\beta) s_i^{\alpha\beta} s_j^{\alpha\beta} - \frac{1}{T} p'(V_\alpha) \sum_{\lambda\mu} p'(V_\lambda \bigwedge H_\mu) s_i^{\lambda\mu} s_j^{\lambda\mu} \tag{10}$$

This expression is used to compute the gradient of the cost function C:

$$\frac{\partial C}{\partial w_{ij}} = -\sum_\alpha \frac{p(V_\alpha)}{p'(V_\alpha)} \frac{\partial p'(V_\alpha)}{\partial w_{ij}} =$$

$$= \frac{1}{T} \sum_\alpha \frac{p(V_\alpha)}{p'(V_\alpha)} \sum_\beta p'(V_\alpha \bigwedge H_\beta) s_i^{\alpha\beta} s_j^{\alpha\beta} - \frac{1}{T} \sum_\alpha \frac{p(V_\alpha)}{p'(V_\alpha)} p'(V_\alpha) \sum_{\lambda\mu} p'(V_\lambda \bigwedge H_\mu) s_i^{\lambda\mu} s_j^{\lambda\mu}$$

$$\tag{11}$$

Now we can write:

$$p(V_\alpha \bigwedge H_\beta) = p(H_\beta \mid V_\alpha) p(V_\alpha)$$

$$p'(V_\alpha \bigwedge H_\beta) = p'(H_\beta \mid V_\alpha) p'(V_\alpha) \tag{12}$$

where

$$p'(H_\beta \mid V_\alpha) = p(H_\beta \mid V_\alpha) \tag{13}$$

because the probability of a hidden state, given some visible state, must be the same in equilibrium whether the visible units were clamped in that state or reached it by free-running. From the previous expression, we obtain:

$$p'(V_\alpha \wedge H_\beta) \frac{p(V_\alpha)}{p'(V_\alpha)} = p(V_\alpha \wedge H_\beta) \tag{14}$$

and since

$$\sum_\alpha p(V_\alpha) = 1 \tag{15}$$

the cost gradient is given by:

$$-\frac{\partial C}{\partial w_{ij}} = \frac{1}{T}[p_{ij} - p'_{ij}] \tag{16}$$

where

$$p_{ij} = \sum_{\lambda\mu} p(V_\lambda \wedge H_\mu) s_i^{\lambda\mu} s_j^{\lambda\mu} \tag{17}$$

and

$$p'_{ij} = \sum_{\lambda\mu} p'(V_\lambda \wedge H_\mu) s_i^{\lambda\mu} s_j^{\lambda\mu} \tag{18}$$

This means that the cost gradient may be computed from the difference of p_{ij}, which is the probability on the training set that neurons i and j are in the same state 1 when the output neurons are clamped, and p'_{ij}, which is the probability for the unclamped situation. The cost gradient function allows to update a weight (w_{ij}) by considering only the probabilities of the states of the neurons i and j. The quantities p_i and p'_j are the mean correlation between the states of the neuron i and j. So that, the weight updating rule is the application of a Hebbian and an anti-Hebbian rule.

The gradient descent technique does not guarantee the global minimum finding. It seems, and it is confirmed by the examples reported in the paper, that the training procedure, described above, determines the global minimum of C thanks to the noise the error introduced in the evaluation of p_{ij} and p'_{ij} lets in. This is equivalent to

$$\Delta W_{ij} = -\eta \frac{\partial C}{\partial W_{ij}} + noise \qquad (19)$$

It has been shown [6][16] that in this condition the global minimum of C is surely reached.

When the probability of a visible state is zero the corresponding term of the cost function becomes zero and this result can be learned only with some weights with an infinite value. For this reason it may be wise to introduce some noise in the input pattern or in alternative to decrement the weights with a given probability [7].

With regard to the representation problem, that is, to the ability of the BM to represent any input-output relation, it is possible to show that the boolean mapping $(0, 1)^{NI} \rightarrow (0, 1)$ can be represented by a three-layer-network where the hidden layer has no more than NP neurons, where NP is the minimum between the number of input patterns producing output 1. Each neuron of a layer is only connected to all the neurons belonging to adjacent layers, no connections are assumed between neurons belonging to the same layer. This means that the global number of weights and biases is

$$NW = NI * NH + 2 * NH + 1 \tag{20}$$

The biases of the input neurons are not considered because the inputs are always clamped.

These results are similar to those obtained for the back propagation model with digital inputs and output [8].

The proof starts with the consideration that the number of patterns that can be obtained by using NI boolean variables is 2^{NI} so the hidden layer must be composed of no more than 2^{NI} neurons.

The k-th hidden neuron (whose state is X_k) corresponds to the k-th input pattern: the weights connecting the input neurons to the k-th neuron are $+\alpha$ $(-\alpha)$, with $\alpha > 2$, if the state of the corresponding input neuron for the k pattern has value 1 (0). N_1^k and N_0^k are the number of (1) and (0) respectively contained in k-th pattern. The bias of the k-th hidden neuron is $(-N_1^k + .5) * \alpha$, while the weight connecting it to the output neuron is $(+1)$ or (-1) if the value of the output state has to be (1) or (0) respectively.

The bias of the output neuron is (-.5) and its state is Y.

In the test phase, when the k-th pattern is presented to the inputs of the BM, it will be shown that the following results are obtained:

- the state of the k-th hidden neuron becomes (1)

- the state the other hidden neurons becomes (0)

- the state of the output neuron becomes (1) or (0) if the weight between it and the k-th hidden neuron is (+1) or (-1) respectively.

When the k-th pattern is presented at the input of the BM the energy function can be expressed as the sum of 3 terms:

$$E(X_k, Y) = E_k + \Sigma_{h \# k} E_h + E_{out} \tag{21}$$

where E_{out} is the contribution due to the output neuron state Y; E_k and E_h are the contributions of the k-th and the h-th hidden neuron.

The mutual terms due to the hidden neurons states and the output neuron state are contained in E_{out} because it will be shown that they have no effect in the computation of X_k and X_h. The values of X_k, X_h and Y, E has a global minimum for, can be obtained by separately computing the global minimum of E_k, E_h and E_{out}.

Consider E_k:

$$E_k = -\alpha N_1^k X_k - (-N_1^k + 0.5)\alpha X_k = -0.5\alpha X_k \tag{22}$$

which is minimum for $X_k = 1$. Since $\alpha > 2$ the mutual term between X_k and Y:

43

$$\pm X_k Y$$

has no influence on the determination of X_k.

Consider E_h:

$$E_h = -\alpha M_1^h X_h - (-\alpha)(N_0^h - M_0^h)X_h - (-N_1^h + 0.5)\alpha X_h \qquad (23)$$

where M_1^h and M_0^h are the number of inputs with value 1 and 0 for which the weight between the input neurons and the h-th neuron is $(+\alpha)$ and $(-\alpha)$ respectively.

Since

$$M_1^h \le N_1^h \, and \, M_o^h \le N_o^h \qquad (24)$$

and

$$D = (N_1^h - M_1^h) + (N_o^h - M_o^h) \ge 1 \qquad (25)$$

(it is impossible by hypothesis that $M_1^h = N_1^h$ and $M_o^h = N_o^h$ for the h-th neuron), we obtain

$$E_h = (D - 0.5)\alpha X_h \qquad (26)$$

The minimum of E_h is obtained for $X_h = 0$ and, since $\alpha > 2$, the mutual term between X_h and Y has no influence in the determination of X_h.

Finally consider E_{out}:

$$E_{out} = -(\pm 1)X_k Y - \Sigma_{h \neq k}(\pm 1)X_h Y - (-0.5)Y \qquad (27)$$

Since

$$X_i = \delta_{ik} \qquad (28)$$

$$E_{out} = [-(\pm 1) + 0.5]Y \qquad (29)$$

The minimum of E_{out} is obtained for Y=1 if the weight of the connection between X_k and Y is (+1) and Y=0 if the weight is (-1).

This is the required output for the k-th pattern.

The number of neurons in the hidden layer can be reduced from 2^{NI} to NP where NP is the minimum between the number of input patterns producing output 0 (case a) and the number of input patterns producing output 1 (case b).

In the case (a) the network we have previously described must be modified in the following way:

- The weights of the connections between hidden neurons and the output neuron have value (-1)

- The bias of the output neuron is (0.5)

In the case (b):

- The weights of the connections between hidden neurons and the output have value (+1)

- The bias of the output neuron is (-0.5)

The representation theorem, just presented, does not give any indication about the minimum number of weights which are necessary to solve a specific problem.

As an example, consider the NI-parity problem [9] requiring the output of the neural network to be 1 or 0 if the number of 1 in the input pattern is even or odd respectively.

It can be shown that this problem can be solved by using one hidden layer with NH=NI, while NP is equal to 2^{NI-1}, resulting in a strong reduction of NW .

In fig. 1 the solution is shown: it is derived from the same one for MLP [9].

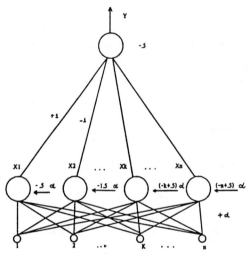

Fig. 1 - The solution for the n-parity problem.

All the weights connecting each input to each hidden neurons have the same value ($\alpha > 2$). The biases of the hidden neurons are different: the bias value of the k-th hidden neuron is (-k+0.5). The connection weight between the k-th hidden neuron and the output neuron is (+1) or (-1) if k is odd or even respectively. The bias of the output neuron is -0.5.

If the number of (1) in the input pattern is N_1, the energy related to the k-th hidden neuron is:

$$E_k = -\alpha N_1 X_k - (-k + 0.5)\alpha X_k = -(N_1 - k + 0.5)\alpha X_k \qquad (30)$$

The value of X_k is (0) or (1) if $N_1 < k$ or $N_1 >= k$ respectively and since $\alpha > 2$ the mutual term $-((-1)^{k+1})X_n Y$ has no influence.

The energy terms related to the output of the network are:

$$E_y = -\beta Y - (-0.5)Y \qquad (31)$$

where $\beta = 1$ or 0 if the number of hidden neurons whose state is 1 is odd or even respectively. The value of the output is Y=1 if $\beta = 1$ and Y=0 if $\beta = 0$.

This is the required output to solve the parity problem.

For specific problems the number of weights can be furtherly reduced by using:

- two or more hidden layers;

- interconnection between non-adjacent layers;

- interconnection between neurons of the same layer.

It must be stressed that the representation theorem or the knowledge of the minimum number of weights do not guarantee both the complete and fast learning.

In fact the complete and fast learning depends on the reliability and on the speed of the particular optimization algorithm used to minimize the cost function.

3   COMPUTER SIMULATION

The BM has been implemented in C language on VAX/VMS and some experiments have been carried on in order to explore the learning capabilities with various topologies. A C-like code is given in the following:

```
for ( i=0; i < number_of_learning_cycles; i++) {
    for ( l = 0; l < 25; l++) {
        for ( k = 0; k < number_of_examples; k++) {
            clamp_inputs_and_output();
            set_random( hidden_neurons);
            equilibrate(hidden_neurons);
            collect_statistics(p);
            clamp_inputs();
            set_random( hidden_neurons + output_neuron);
```

```
        equilibrate(hidden_neurons + output_neuron);
        collect_statistics(p1);
    }
}
    update_weights(p, p1);
}
```

The parity problem with inputs varying from 2 to 4 has been tested in the experiments using a BM with N1=NH. An order of magnitude of improvement has been reached in the learning time by avoiding the equilibrating process in the statistical acquisition of neuron state probabilities while the annealing schedule has been used in the test phase: this is similar to the algorithm in [7].

The effects of modifying the topology have been explored. Starting from a full interconnected BM the pruning of some connection has been evaluated and compared with the full connected network. The goal was to reduce the number of weights to be found in the learning phase in order to simplify the hardware implementation.

Four types of connections have been tested:

a) a full interconnected network.

b) The same of (a) with the pruning of the interconnections among neurons of the same layer.

c) The same of (a) with the pruning of the interconnections between input and output neurons.

d) A network with interconnections among neurons of adjacent layers.

Several tests on temperature T to use have been accomplished.

The employed values are those where more reliable results have been obtained.

For each interconnection type 10 experiments have been carried out.

Each experiment consisted of a maximum of 10.000 learning cycle, and a learning cycle is the presentation of 25 epochs, (an epoch represents all the pattern of the training set).

The values of the percentage of correct identification, shown in table 1, are computed through the procedure described in the following. For each member of the epoch the network is equilibrated, then the average of the output neuron value is computed over 22 times and it is transformed into a binary number, that is, if it is greater than 0.5 it is assumed to be 1, otherwise it is 0. This operation is repeated ten times and the average of the averages is computed and transformed into a binary number. Finally the computed value is compared with the output target. The average of correct response is then computed over every member of the epoch. The percentage of correct identification is the average of the averages of correct response over ten runs of the BM.

The learning speed is the average number of learning cycles useful to reach 90% of the obtained percent of correct identification.

In table 1 some of the obtained results are presented.

TABLE 1

T	NI	Interconnection type	Percent of correct identification	Learning speed
25	2	a	100	500
25	2	b	100	600
25	2	c	100	1100
25	2	d	100	900
25	3	a	96	1500
25	3	b	100	1500
25	3	c	97	3000
25	3	d	99	2500
40	4	a	98	5300
40	4	b	95	5000
40	4	c	84	4600
40	4	d	84	6200

The obtained results stress the importance of the connections between input and output neurons to achieve convergence.

A possible explanation of these results lies on the well-known fact that, in the parity problem, with direct connections between input and output neurons it is possible to use a number of hidden neurons which is the integer greater equal NI/2, then with interconnection types (a) and (b) the network has more possible solutions than in the other cases.

On the contrary, connections among hidden neurons are much less important or in some instances they appear unfavourable.

This is more difficult to explain but perhaps a possible suggestion is that no topology is known where interconnections among hidden neurons are successfully used.

As a general comment the learning speed obtained with BM has been found satisfactory in comparison to the back-propagation results [1].

BM can be extimated to be 3 times slower than back-propagation.

4   CONCLUSION

In this paper a computer simulation of BM has been presented.

And it has been shown that in a hard learning problem such as the n-parity the learning abilities of BM are satisfactory.

These results must be considered together with the advantages of BM related to its hardware implementation. Moreover BM is based on solid theoretical concepts originating from statistical mechanics.

5   REFERENCES

[1] D.E. Rumelhart, J.L. McClelland (Eds.): Parallel distributed processing: explorations in the microstructures of cognition. Vol.1, MIT PRESS, Cambridge, Mass., 1986.

[2] G.E. Hinton, T.J. Sejnowski: Learning and relearning in Boltzmann

Machines. In [1], p. 282-317.

[3] D.H. Ackley, G.E. Hinton, T.J.Sejnowski: A learning algorithm for Boltzmann Machines. Cognitive Science, Vol. 9, 1985, p. 147-169.

[4] M. Minsky, S. Papert: Perceptrons. MIT PRESS, Cambridge, Mass., 1988.

[5] S. Kirkpatrick, C. Gelatt, M. Vecchi: Optimization by simulated annealing. Science, Vol. 220, 1983, p. 671-680.

[6] S. Geman, C.R. Hwang: Diffusions for global optimization. SIAM J. Control and Optimization, Vol. 24, no. 5, September 1986, p. 1031-1043.

[7] T.J. Sejnowski, P.K. Kienker, G.E. Hinton: Learning symmetry groups with hidden units: beyond the perceptron. Physica 22D, 1986, p. 260-275.

[8] B. Moore, T. Poggio: Representation properties of multilayer feed-forward networks. Abstracts of the first annual INNS meeting, Boston, 1988, Neural Networks, Vol. 1, Supplement 1, 1988, p. 203.

[9] D.E. Rumelhart, G.E. Hinton, R.J. Williams: Learning internal representations by error propagation. In [1], p. 330-335.

[10] Metropolis, N.,et al: Equation of State Calculations by fast computing machines. Journal of Chemical Physics 21, 1953, p.1087-1092.

[11] Kullback, S: Information Theory and Statistics. Wiley and Sons, New York, N.Y. 1959.

[12] Aarts, R.H.L., Jan J.M. Korst: Computations in Massively Parallel Networks Based on the Boltzmann Machine: A Review. Parallel Computing 9, 1989, p. 129-145.

[13] Aarts, E.H.L., Jan H.M. Korst: Simulated Annealing and Boltz-mann Machines: A Stochastic Approach to Combinatorial Optimization and Neural Computing. John Wiley and Sons, New York, N.Y., 1989.

[14] Van Laarhoven, P.J.M., E.H.L. Aarts: Simulated Annealing: Theory and Application. Reidel Publishing Company, Boston, MA, 1987.

[15] Van Laarhoven, P.J.M.: Theoretical and Computational Aspects of Simulated Annealing. CWI Tracts, Amsterdam, The Netherlands, 1989.

[16] Geman, S, D. Geman: Stochastic Relaxation, Gibbs Distributions and the Bayesian Restoration of Images. IEEE Transactions of Pattern Analysis and Machine Intelligence, 6, 1984, p. 721-741.

3

A THEORETICAL ANALYSIS
OF POPULATION CODING
IN MOTOR CORTEX

Terence D. Sanger

MIT, E25-534
Cambridge, MA 02139

ABSTRACT

I consider some theoretical issues surrounding recent evidence of population coding of control variables in motor cortex. The mathematical formalism used is that of radial basis functions, and some of the interesting properties of these functions are discussed. It is shown that a distributed representation based on radial basis functions allows the use of linear techniques to approximate certain nonlinear output functions or coordinate changes. The significance of this for interpreting population coding in motor cortex is discussed. It is shown that the experimentally determined representation depends on both the input and output coordinate systems, as well as on the task being performed and the way in which the animal chooses to control that task. I describe the relation between the theoretical predictions and the biological data, and I show that certain results not predicted by theory may represent constraints on the biology.

1 INTRODUCTION

If we consider the brain as a device for performing prescribed actions on the environment, then it must have the ability to translate from the form in which tasks are specified into a form which describes external actions (Lacquaniti, 1989). Presumably, an organism has high-level goals which are specified in terms of desired outcomes such as "find food", or "run away from tigers". Such goals must be translated into sets of commands which can be sent to muscles to actually achieve the goal. To the extent that animals survive in the jungle, we can assume that biological brains are capable of solving this problem.

As a first step toward understanding how this is accomplished, it may be useful to ask whether there exist intermediate representations which are on the path between task representations and motor representations. If such representations exist, their form may not be determined by the task at hand, but might perhaps be determined by the computational requirements of a neural system.

In this paper, I will investigate representations of proximal arm reaching in primate motor cortex. Since there are cells of motor cortex which synapse directly onto alpha motoneurons, one might expect that the representation in this part of the brain would be closely linked to the motor representation in terms of the motoneuron outputs. Knowing at least the function, if not the form, of the output of motor cortex might allow us to make some simple claims about the possible types of representations.

There has recently been much important work suggesting that the motor cortical representation is a "population code" (Kalaska et al., 1989, Georgopoulos et al., 1989b, Georgopoulos et al., 1989a, Georgopoulos, 1988, Schwartz et al., 1988, Georgopoulos et al., 1988, Kettner et al., 1988, Georgopoulos, 1986, Georgopoulos et al., 1985, Georgopoulos et al., 1983, Georgopoulos et al., 1982). The idea of population coding is that values are not represented by single neurons, but rather by the combined activity of an entire group of neurons. Very often, the actual firing rates of the neurons will be considered much less significant than the pattern of firing over the population. In a sense, one can think of population coding as a "digital" representation of "analog" data. In an analog representation, a single value is represented directly by a single physical quantity, such as a neuronal average firing rate. In a digital representation, a single value is represented by a pattern of bits encoded in multiple transistors. Many useful properties arise from a digital representation, including insensitivity to noise and ease of computation. The main disadvantage is that the amount of required hardware increases; we may have to use 16 transistors to represent a value which could be stored in analog format by only one.

Very often, when people refer to population coding, they mean a particular type of population code which is sometimes known as "coarse coding" (Hinton et al., 1986). Unlike the example of a digital representation in which a given bit may be "on" for many different values of the data, in coarse coding each coding element (neuron) gives information about only a local region of similar data values. For example, one neuron might fire whenever the arm moves left, while another would fire if the arm moves forward. A large number of neurons would be able to cover all possible directions. The question then arises as to whether two (or more) neurons would fire if the arm moves diagonally left and forward. This can be considered a measure of the "coarseness" of the coding. We often speak of neurons as having motor "receptive fields" which might describe the range of directions of motion for which that neuron responds.

There are many examples of population coding in regions other than motor cortex. Coarse coding has been found in areas 5 (Kalaska, 1988, Kalaska *et al.*, 1983), and 7 (Andersen and Zipser, 1988, Steinmetz *et al.*, 1987, Andersen *et al.*, 1985) of Parietal cortex, in the superior colliculus (van Gisbergen *et al.*, 1987, McIlwain, 1975), and in the visual and auditory systems (Knudsen *et al.*, 1987, for review). It is tempting to speculate that population coding may allow exchange of information between these multiple modalities (Knudsen *et al.*, 1987).

In this paper, I will use the mathematical formalism of Radial Basis Functions (RBFs) (Powell, 1987, for review) to model biological coarse coding. There has been considerable interest recently within the robotics community in the possible use of this representation for adaptive control of robots. In section 2 I will discuss some of the mathematical implications of this model. In sections 3 and 4 I will discuss the relation of the radial basis function model to the experimental data from primate motor cortex.

2 RADIAL BASIS FUNCTIONS

Radial basis functions (RBFs) have been successfully used to solve function approximation problems in several domains (Farmer and Sidorowich, 1989, Renals and Rohwer, 1989b, Renals and Rohwer, 1989a, Saund, 1989, Broomhead and Lowe, 1988, Moody and Darken, 1988). Reviews of the theory can be found in (Poggio and Girosi, 1990, Powell, 1987, Klopfenstein and Sverdlove, 1983). In this section I will briefly summarize the formulation of RBFs, and I will relate the well-known results in this field to the problem of motor coordinate transformations.

2.1 CURVE FITTING AND FUNCTION APPROXIMATION

Many problems in robotics or biological motor control can be formulated as "function approximation" problems. Such problems try to find a function $f(x)$ from finitely many samples of input-output pairs (x_t, y_t) where $x_t \in R^n, y_t \in R^m$. For any given finite set of samples, there are infinitely many continuous functions which compute $y_t = f(x_t)$ $\forall t$ on that set. In order to constrain the problem somewhat, one can pick a finite-dimensional space of functions and attempt to find a function $\hat{f}$ in this space which approximately computes $y_t \approx \hat{f}(x_t)$. Although there are many possible choices of finite-dimensional space, a particularly convenient choice is to pick a Hilbert space, in which functions may be written

$$\hat{f}(x) = \sum_{i=1}^{N} c_i \phi_i(x) \tag{1}$$

where $\{c_i\}$ is a set of constants, and $\{\phi_i\}$ is a finite set of N scalar-valued basis functions (see figure 1. (I will write $\hat{f}$ as a scalar-valued function, with the implicit assumption that we can make vector-valued functions by combining multiple scalar-valued ones.) Since this is a finite-dimensional space, it is nowhere dense in the set of all possible functions, so we should not expect to be able to approximate arbitrary functions with small error. However, if we know that the function f is likely to lie within a certain region of function space, then

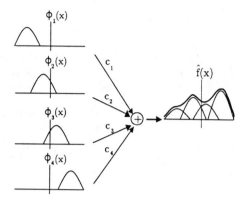

Figure 1: Approximation of a function $f(x)$ using a superposition of fixed basis functions.

we may be able to choose a basis $\{\phi_i\}$ which gives good approximations within this region. (See (Sanger, 1990a) for one method of choosing the basis functions.)

Once we have chosen the basis functions, finding the constants c_i is a linear problem. If we write all the outputs as a row vector $Y = [y_1, \ldots, y_T]$ and the basis function values as a matrix $A_{it} = \phi_i(x_t)$ then we have the linear equation

$$Y \approx CA$$

where C is a row vector of constants c_i. If A is full rank (meaning that the basis function values $\phi_i(x_t)$ do not lie on a low-dimensional surface in R^N), then the choice of C which minimizes the mean-squared error in Y is given by

$$C = YA^\dagger$$

where $A^\dagger$ is the Moore-Penrose pseudo-inverse of A (Penrose, 1955).

Note that we have now reduced the nonlinear surface fitting problem to a linear problem in matrix algebra (see also (Broomhead and Lowe, 1988)). This was accomplished by making several simplifications:

1. We assumed that a function f exists such that $y_t = f(x_t)$.

2. We attempt to approximate f by $\hat{f}$ where $\hat{f}$ lies within a finite-dimensional space.

3. We restrict our finite-dimensional space to be a linear vector (Hilbert) space.

4. We assume that we can choose basis functions ϕ_i such that all likely possibilities for f are reasonably close to the span of $\{\phi_i\}$.

5. We choose the constants c_i to minimize the mean squared error of the output values $\sum \|y_t - \hat{f}(x_t)\|^2$

The technique of radial basis functions and the idea of population coding rest strongly upon these assumptions.

An example of the use of basis functions for motor control is found in (Kawato *et al.*, 1987). Kawato *et al.* used products of trigonometric functions of the joint positions, velocities, and accelerations as the basis functions, and computed the joint torques required to produce the given joint trajectories as a linear combination of the basis functions. Another example is found in (Ritter *et al.*, 1989). Ritter *et al.* use basis functions derived from the Kohonen algorithm (Kohonen, 1982, Grossberg, 1976) which look like non-intersecting polygonal regions in joint space whose centers can be moved by learning. For each region in joint space specified by a basis function $\phi_i = h(\theta - \theta_i)$, the output of the network is given by $c_i = J(\theta_i)$ where θ is the current joint position, θ_i is the center of the polygonal response region, $h()$ is a function which is 1 for values of its argument near zero, and $J(\theta_i)$ is the Jacobian matrix at θ_i which relates joint coordinates to external coordinates. The network thus approximates the nonlinear function $J(\theta)$, and this output is used to compute the correct torques to apply to achieve a desired trajectory.

2.2 RADIAL BASIS FUNCTIONS

Radial basis functions are a particular choice of basis ϕ_i. They can be described by

$$\phi_i(x) = h(\|x - \xi_i\|) \tag{2}$$

where h is a scalar-valued function. The name "radial basis function" comes from the fact that the value of $\phi_i(x)$ depends only on the radial distance of x from the center value ξ_i. If h is monotone decreasing, then the basis has the intuitively nice property of being local in the sense that ϕ_i will have its largest values in a region near ξ_i. If ϕ_i approaches 0 for x far from ξ_i, then the choice of weight c_i in equation (1) will depend only on values x_t which lie near ξ_i. This is a nice property computationally, since the matrix A will be sparse, and may thus be much easier to invert.

In general, the choice of the centers ξ_i is arbitrary, and may be chosen to lie either at data points, on a fixed lattice, or near clusters of data. Here, I will use basis functions centered on a fixed lattice. In addition, I will sometimes choose the function h to be a Gaussian

$$h(\|x - \xi_i\|) = \frac{1}{\sigma\sqrt{2\pi}} e^{-\|x-\xi_i\|^2/2\sigma^2}$$

This is not always necessary and most monotone decreasing positive functions will have all the properties described below, but for concreteness sake I will assume we are using Gaussian RBFs. Examples of the use of radial basis functions of this form may be found in (Broomhead and Lowe, 1988, Moody and Darken, 1988, Klopfenstein and Sverdlove, 1983).

2.3 FUNCTION APPROXIMATION

In view of simplification 4 above, we can ask what set of functions f can be well approximated by radial basis functions. Poggio and Girosi (1989) have shown that RBFs are equivalent to generalized splines, and that both representations can be derived from regularization conditions involving smoothness constraints. Thus RBFs represent smooth functions.

For the case of Gaussian RBFs centered on a regular lattice, I can provide a different explanation. In this case, we can consider the outputs $\phi_i(x)$ to be samples of a Gaussian smoothed version of the input space R^n. Viewed as a filtering operation, this is a form of subsampling of the input space, and can therefore represent "frequencies" up to the sampling Nyquist rate. (This filtering is occurring in "value" space rather than time, and was originally suggested in this context in (Klopfenstein and Sverdlove, 1983).) Notationally, we can write:

$$\phi(\xi_i, x) = \int h(\|x' - \xi_i\|)\delta(x - x')dx'$$

so that we can view ϕ as a function of either x or ξ_i. To find the value-frequency response of ϕ, we can Fourier transform with respect to x to obtain:

$$\mathcal{F}[\phi](\xi_i, \omega) = H(\omega)e^{\iota \omega^T \xi_i}$$

where ω is a vector of value-frequencies. The approximation of our desired function $f(x)$ is now given in the value-frequency domain by:

$$\hat{F}(\omega) = \sum c_i H(\omega)e^{\iota \omega^T \xi_i}.$$

If ξ_i is evenly spaced with spacing ϵ along the direction of a given unit vector r, then we can write $\xi_i = i\epsilon r$ and the sampling frequency is $\omega_s = 2\pi/\epsilon$, giving:

$$\hat{F}(\omega) = H(\omega)\sum c_i e^{2\pi\iota\omega^T ri/\omega_s}. \tag{3}$$

(The right hand side should be summed over all possible lattice directions r.) The power spectrum is given by:

$$\hat{F}^2(\omega) = H^2(\omega)V(\omega) \tag{4}$$

where $V(\omega)$ is a real scaling term which depends on the values of the c_i's. In the value-Fourier domain, the approximation problem becomes what is the range of possible spectra which $\hat{F}$ can have, and is this a good approximation to the actual spectra F which occur? Note that there is an implicit requirement that H must be small whenever ω is greater than $\omega_s/2$ in order to satisfy the Nyquist sampling theorem. In addition, H must be a radial function since h is radial. This means that H will function as a radially-symmetric low-pass filter. Equations (3) and (4) then give the following restriction on the set of functions $\hat{F}$: the power spectra will be low-pass (functions $\hat{f}$ will be smooth), and will have low energy in frequency bands for which $H^2(\omega)$ is small.

There is thus an interaction between the sampling rate ω_s, the shape of the filters h, and the functions $\hat{F}$ which can be approximated. ω_s determines the maximum frequency to which H should respond, and this in turn determines the maximum frequency which will be present in $\hat{F}$. In addition, if H is small for frequencies below ω_s, then these frequencies will not be able to be well-represented in $\hat{F}$.

In choosing a set of basis functions, it is important to understand these relationships so that the basis functions are (1) neither too far apart (low ω_s) and (2) do not have missing frequencies (zeros in $H(\omega)$ for $\omega < \omega_s$). A failure of the first type occurs if we choose

$$h(\|x\|) = e^{-\|x\|^2/2}$$

and spacing $\epsilon = 1$. A failure of the second type occurs if we choose

$$h(\|x\|) = 1 + \cos(\omega\|x\|)$$

(with the argument restricted to $(-\pi, \pi)$ so that h is "local"). In the first case we will introduce spurious high-frequency components into $\hat{f}$, and in the second, we will be unable to represent frequencies other than ω.

From the discussion above, we conclude that a function f can only be approximated if it is "sufficiently smooth" in the sense that it does not have frequency components outside the response range of the basis functions themselves. Even if the functions are smooth, that does not guarantee that a good approximation is possible. For instance, if the basis functions are local but are spaced infrequently, then it will be impossible to approximate a constant function well. In general, good approximation of arbitrary functions depends on the number of basis functions N as well as the power spectrum of the functions to be approximated.

2.4   DISTRIBUTED REPRESENTATIONS

So far I have discussed the problem of using radial basis functions to approximate a function

$$f : x \in R^n \mapsto y \in R^m$$

However, it is also possible to consider RBFs as defining a mapping

$$\Phi : x \in R^n \mapsto \chi \in R^N$$

where χ can be thought of as the column vector $(\phi_1(x), \ldots, \phi_N(x))^T$ (see figure 2). In neural network literature this vector is often called the "hidden layer". The function approximation problem on R^N is the problem of finding a linear functional

$$C : \chi \in R^N \mapsto \hat{y} \in R^m$$

such that

$$\hat{y} = \hat{f}(x) \approx y = f(x)$$

Here I will discuss some properties of the intermediate representation χ.

Since x lives in an n-dimensional space, all values of $\chi = \Phi(x)$ will live on an n-dimensional smooth manifold in R^N (assuming certain smoothness conditions on the ϕ_i). Any particular value of x will be represented by a unique point on the manifold. The coordinates of this point in R^N will in general be described by N numbers. Although it may be possible to determine the point x uniquely with fewer than N coordinates, in general this will not be possible (for example, it requires three coordinates to specify the position of a point on a 2-sphere in R^3). This is how I have chosen to represent the idea of population coding. I have taken n values and represented them using $N \gg n$ values. As an example, consider representing a single number x using $\phi_i(x) = \cos(\omega_i x)$. If $x \in (-\pi, \pi)$ and $w_i < 1$, then x can be uniquely determined from any single value of $\phi_i(x)$ by using the arccos function. However, I have encoded x in a much higher dimensional space.

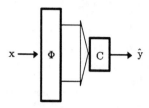

Figure 2: Network structure with distributed "hidden layer" $\chi = \Phi(x)$ and linear output layer C.

The computational usefulness of the encoding comes from the fact that the original space R^n is "flat", while the population coded manifold may be "curved" in R^N. To understand this, I now need to look at how the output y is computed from χ using the linear functional C. C can be thought of as a projection operator. In other words, for the scalar-output case it is a set of N numbers which forms a vector in R^N, and the value of $C\chi$ is the one-dimensional projection of the value χ onto the vector C. The function $\hat{f} = C\Phi(x)$ can be thought of as mapping x onto a curved manifold and then looking at the C-component of points in this manifold. We saw in the last section that this allows approximation of any function which has certain smoothness properties determined by the representation $\Phi(x)$.

Now, imagine that we do not population code, so that $\Phi : R^n \to R^n$ maps x onto itself. Then C is an n-dimensional vector, and χ can live anywhere in the space R^n. The set of functions which we can compute now consists of the linear functions given by the linear functional C. We have thus lost the ability to approximate smooth nonlinear functions. Note that we get pretty much the same effect if Φ does map into a higher-dimensional space R^N, but that the manifold in R^N is flat. For instance, if $\phi_i(x)$ is linear, then we can write

$$\phi_i(x) = p_i^T x,$$

$$\Phi(x) = Px$$

where p_i is a vector in R^n and P is an $N \times n$ matrix. Then $C\Phi(x) = CPx$ is the conjunction of two linear functions and thus linear, so once again we cannot approximate nonlinear functions. Of course, this example does not use strictly radial basis functions, but it illustrates the idea that the choice of distributed representation can affect the possible functions which can be computed.

2.5 COORDINATE TRANSFORMATIONS

For a different choice of N basis functions $\{\phi_i'\}$, x will map into a different point $\chi' = \Phi'(x)$. What is the relationship between the distributed representations χ and χ'? Let g be a map such that $\Phi'(x) = g(\Phi(x))$. Then if g is sufficiently smooth in the sense of section 2.3, we can approximate each dimension of it using the standard radial basis function approximation. In other words, if $g(\Phi(x)) = (g_1(\Phi(x)), \ldots, g_N(\Phi(x)))^T$, then we can approximate

$$g_i(\Phi(x)) \approx B_i\Phi(x), \text{or}$$

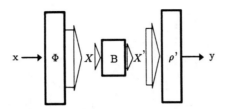

Figure 3: Linear transformation B mapping distributed representation $\chi = \Phi(x)$ onto $\chi' = \Phi'(x)$ before computing the nonlinear output function ρ'.

$$g_i(\chi) \approx B_i \chi$$

with B_i a row vector, as we did before. Now, if we let B be an $N \times N$ matrix whose rows are the B_i's, then we have

$$\chi'(x) = g(\chi(x)) \approx B\chi(x)$$

and we see that the relationship between χ' and χ may be approximately linear. We can compute the matrix B using the pseudoinverse:

$$B = \Upsilon'\Upsilon^\dagger$$

where Υ is the matrix whose columns are $\Phi(x_t)$ for different samples x_t, and Υ' is the matrix whose columns are $\Phi'(x_t)$ (see figure 3). Thus any sufficiently smooth nonlinear change of variables in a distributed representation can be approximated by a linear change of variables.

This is true even when the input variables x are different. For example, suppose that x' is proprioceptive information from the arm, while x is visual information about arm position. x' will most likely be in terms of joint coordinates, while x may be in spatial or retinal coordinates. Assume that there exists a smooth map q such that $x' = q(x)$. When we compute the distributed representations $\chi = \Phi(x)$ and $\chi' = \Phi'(x')$ we find that they may still be approximately related by a linear transformation $\chi' \approx B\chi$. We thus have a method for converting visual coordinates to motor coordinates.

We can pseudo-invert the linear mapping between distributions, even if the actual nonlinear mapping g is not invertible. If $\chi' \approx B\chi$, then the least squares linear mapping B' relating the distributed representations in the inverse direction is given by

$$B' = \Upsilon\Upsilon'^\dagger = (\Upsilon'\Upsilon^\dagger)^\dagger = B^\dagger$$

so that

$$\chi(x) \approx B'\chi'(x) = B^\dagger\chi'(x).$$

So, for example, we can convert motor to visual coordinates, or motor and visual coordinates to an intermediate representation which combines information from both, although the fact that $B^\dagger$ is not a true inverse means that some information may be lost along the way (see figure 4).

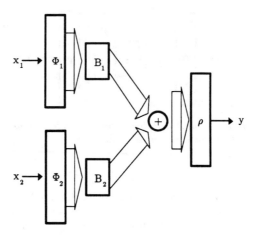

Figure 4: Distributed representations $\chi_1 = \Phi_1(x)$ *and* $\chi_2 = \Phi_2(x)$ *are converted to an intermediate representation and then combined to produce the output value* y.

If the rank of $B^\dagger B$ is the same as the rank of $BB^\dagger$ and is equal to n, then we can approximate any output function $y = f(x)$ equally well by $y = C\chi(x)$ or $y = C'\chi'(x)$ where $\chi'(x) = B\chi(x)$. This gives $C' = CB^\dagger$. This means that neither distributed representation has a significant advantage in terms of output functions, so long as no information is lost when we pass between them. We can thus compute the same output as a linear function of different distributed representations.

Similarly, we can linearly compute many different smooth nonlinear functions from the same representation. This fact has an important implication for interpreting the structure of a distributed representation. Any function f may be approximately linearly related to the elements of the distributed representation. But any other function f' (which depends smoothly on f) will also be linearly related to the elements of the same distributed representation, although through a different matrix C'. We thus cannot draw any conclusions from the existence of a linear mapping from a distributed representation to a measured variable. The output mapping C will depend on both the representation $\Phi(x)$ and on the particular function f which we are trying to approximate. One cannot say that such a representation "encodes some property X", since it clearly also encodes any other property Y which depends smoothly on X.

Suppose an experimenter is investigating a system which has an intermediate layer which is a distributed representation. We can write such a system as

$$f(x) = \rho(\Phi(x))$$

where $\rho : R^N \to R^m$ is some nonlinear function. Then if f is sufficiently smooth in terms of Φ, there will exist a linear mapping C such that

$$f(x) \approx C\Phi(x).$$

Essentially, this means that for this particular representation $\Phi(x)$, the output map ρ appears to be linear. In fact, ρ may not be linear, and we must realize that the linear approximation C depends both on f and on $\Phi(x)$. The fact that ρ appears to be linear is caused by the distributed representation $\Phi(x)$. The experimenter will always find an approximately linear relationship between elements of this distributed representation and the output, but will not be able to draw any conclusions about the function ρ from this information.

2.6 SUMMARY

So far, I have shown that distributed representations possess some useful computational properties. They convert nonlinear problems into linear problems, and allow certain smooth nonlinear mappings to be approximated. A nonlinear coordinate change in a distributed representation can under certain conditions be approximated by a linear transformation. This means that any given representation can be thought of as encoding information simultaneously in several different coordinate systems which are related through smooth invertible mappings. Any distributed representation will be approximately linearly related to any other (smoothly dependent) distributed representation.

An additional property that distributed representations can have is redundant encoding of values. This allows a greater tolerance for processing noise in the distributed representation itself. Intuitively, we might expect that if many elements of a representation χ are correlated, then the output matrix C will depend on all of them and be able to reject isolated noise in any one. However, this is true only in very specific circumstances, and it is difficult to make general statements about the noise properties of distributed representations.

Distributed representations have been used successfully in a variety of practical applications, such as are found in (Moody, 1989, Renals and Rohwer, 1989a, Renals and Rohwer, 1989b, Moody and Darken, 1989, Moody and Darken, 1988, Raibert, 1978). Much of the interest in these representations comes from studies of learning algorithms which can be used either to pick a good representation (usually by moving the basis function centers ξ_i) or a good output map C (Linsker, 1989, Ritter *et al.*, 1989, Ritter and Schulten, 1988a, Kohonen, 1988, Ritter and Schulten, 1988b, Miller *et al.*, 1987, Kohonen, 1982, Raibert, 1978, Grossberg, 1976). Systems incorporating coarse coding are able to learn quickly to approximate maps from samples of inputs and outputs presented sequentially. I will not consider learning algorithms here, although this is an important topic and has motivated much of the study of radial basis functions.

3 CONTROL OF DISTRIBUTED REPRESENTATIONS

Up to this point, I have investigated the problem of finding an output matrix C given a particular distributed representation $\Phi(x)$ and a desired function $f(x)$. Now I will look at a different problem, in which we are given a function $\rho : \chi \in R^N \mapsto y \in R^m$ and we want to find a distributed representation $\Phi(x)$ such that

$$f(x) \approx \rho(\Phi(x)).$$

This is an attempt to model the problem of how, given a function ρ which takes activation levels at motor cortex and produces muscle activity, we can choose an appropriate activation pattern $\Phi(x)$ for the cortical cells which accomplishes a desired task (see figure 2). Clearly, this is a problem solved by all behaving animals, and we here look at some constraints on the way that this task can be performed.

In general $N \gg m$ (above I often assumed that $m = 1$, but the results generalize easily to more outputs), which means that for almost all output values $y \in R^m$ there is an $N - m$ dimensional subspace of R^N all of whose points map to y. For a given set of values which we would like as outputs, there is an infinite number of ways of choosing a distributed representation. This ambiguity can be useful. Not all representations will work, however, and the choice available will depend on ρ (Knudsen et $al.$, 1987, page 57).

In choosing a distributed representation, we can choose one which is easy to compute, or one which helps reduce noise, or one which optimizes some chosen criterion. It is possible for two different representations to coexist in the sense that both may produce acceptable values of the outputs, yet they may not be easily computable from each other. For instance, if $\chi = \Phi(x)$ is one representation such that $f(x) \approx \rho(\Phi(x))$, and $\chi' = \Phi'(x)$ is another with $f(x) \approx \rho(\Phi'(x))$, there may not be any sufficiently smooth invertible mapping between χ and χ' that we can find a linear function which approximates it. Nevertheless, both representations solve the task. Which one is used at any given time is a matter of convenience.

In this context, we can think of any particular distributed representation as being a "control strategy". The function $\Phi(x)$ specifies how we control the representation for each x. Biological systems must discover control strategies to solve particular problems. (This appears to be difficult, until we realize that there will often exist a linear mapping between a bad control strategy Φ' and a good control strategy Φ, as explained in section 2.) To determine the control strategy used by a biological system, we examine the values of the elements of $\chi_i = \phi_i(x)$. We must choose a coordinate system for x, and this will determine the form of the ϕ's. Usually, x will be defined in some set of "task coordinates" defined by the current goal of the network. For any given task which is completely specified by stimulus parameters x, $\phi_i(x)$ will be found to be smoothly dependent on only x.

Note that we do not learn anything about the output mapping ρ, or the possible existence of other control strategies which are used in other situations. It is possible, however, to learn about the linear approximation to the output mapping. Since this approximation depends on the distributed representation, we will write it here as C_Φ.

As a concrete example, assume that motor cortex can be considered a distributed representation. Then an extracellular electrode recording activity in a behaving animal gives information about the control strategy being used by the animal to accomplish a particular task. The linear approximation C_Φ to the output map ρ can be investigated using the naturally occurring patterns of activity $\Phi(x)$ (the animal's control strategy), so long as the patterns span R^N (this is a strong constraint on the patterns, and will not in general be satisfied). To do this, measure the activity χ_t and the outputs y_t where the output space could be joint angles, hand position, or any other variable of interest. Form the matrices Y and Υ whose columns are the vectors y_t and $\chi_t = \Phi(x_t)$. Then we have $Y = C_\Phi \Upsilon$, and

$$C_\Phi = Y \Upsilon^\dagger$$

gives the best approximation to a linear output map.

Since many smooth functions can be approximated using a linear output map, $\hat{y}_t = C_\Phi \chi_t$ may be a good approximation to y_t. This does not mean that the animal actually performs a linear transformation to compute the output, but rather that whatever transformation is performed can be well approximated using a linear map C_Φ, for the particular control strategy $\Phi(x)$. Note that the choice of coordinates for y is arbitrary. For instance, if y represents the Cartesian coordinates of the hand then although this is biologically relevant, these coordinates may never be explicitly computed by the animal. The matrix C_Φ which we obtain will be dependent upon our choice of output coordinate system, and does not necessarily represent any physical structure in the animal.

Note that C cannot be investigated using microstimulation techniques. Stimulating the cortex locally generates a particular choice of representation Φ for which only small local regions of cortex are active. This may not correspond to any physiological control strategy at all, let alone one which is relevant to the task being investigated. Therefore, the matrix C which is found by this technique will not provide useful information. In order to investigate the output function ρ, one would have to stimulate multiple cortical cells in all possible combinations, since in general a nonlinear function like ρ will have interactions between different inputs, and cannot be expected to exhibit linear superposition.

I have been circling the issue here of how one can define the concept of a "motor field" analogously to a sensory "receptive field". Intuitively, the motor field should be the action which occurs when a single cell is stimulated. It would seem that this action should be defined by a single column of the matrix C (assuming all other cells are held at zero). Yet I showed above that C depends on both the control strategy $\Phi(x)$ and the output representation which we have chosen. Therefore, this is not a useful general definition. However, for any particular Φ, we can think of a column of the matrix C_Φ as the contribution which a particular cell makes to the output for this task, when all the other cells are activated appropriately for this control strategy. The cell firing rates and the columns of the matrix will of course be different if a different control strategy is used, even if it produces the same outputs.

If we measure the columns of C_Φ for a particular Φ and call these the motor fields for each cell, we obtain a distribution of points in R^m. The distribution in R^m will depend on our choice of output coordinate system. In some output coordinates, the distribution may be uniform, while in others it may be highly skewed. For example, if the distribution is uniform in hand position, it is not uniform in joint angle.

In summary, measurement of the activation of elements indicates the control strategy being used and the linear approximation to the output map. To determine the full nonlinear output function ρ we would have to stimulate multiple cortical cells in all possible combinations. We can think of these two operations as pseudo-inverses of each other if we view the control strategy Φ as a nonlinear function which assigns a distributed value χ to every output y, and the output function ρ as a non-linear function which assigns an output y to every distributed value χ. It is important to realize that for a particular control strategy, the representation and output map which we infer from our experiments will be determined by the arbitrary choice of input and output coordinate systems, and that no "natural" coordinate system exists. Also, we will not be able to determine the function ρ with any certainty, although for a particular task the linear approximation may be sufficient.

4 DISTRIBUTED REPRESENTATIONS IN MOTOR CORTEX

In this section I will discuss the implications of the results derived above for interpreting data from cortical recording and microstimulation experiments (see also (Sanger, 1990b)). The question which these experiments often seek to address is: "what is the internal representation of motor commands?" This can also be phrased as: "what coordinate system is coded by motor cortex?" Many researchers claim that a particular variable is coded by motor cortex if the firing rate of cortical cells is linearly related to that variable. The claim which I will attempt to support in this section is that with this definition, the motor cortex encodes every variable we look at. In other words, the answer to the question "does the motor cortex encode X?" is almost always "yes!". So the question "what coordinate system is coded" is meaningless, since there is no unique answer.

4.1 EXPERIMENTAL RESULTS

Recently, there have been several studies of the response of motor cortical neurons while an awake monkey executes a reaching task (Kalaska *et al.*, 1989, Georgopoulos *et al.*, 1989a, Georgopoulos *et al.*, 1989b, Georgopoulos, 1988, Kettner *et al.*, 1988, Georgopoulos *et al.*, 1988, Schwartz *et al.*, 1988, Georgopoulos *et al.*, 1985, Georgopoulos *et al.*, 1983, Murphy *et al.*, 1982, Georgopoulos *et al.*, 1982). These studies have produced the following general results:

1. The firing rate of most individual cells is proportional to the cosine of the angle between a "preferred" vector and the direction of hand motion (Schwartz *et al.*, 1988, Georgopoulos *et al.*, 1985, Georgopoulos *et al.*, 1983, Georgopoulos *et al.*, 1982), the static position of the hand (Kettner *et al.*, 1988), or the direction of applied load (Kalaska *et al.*, 1989).

2. The directions of the "preferred" vectors are distributed approximately uniformly over the sphere.

3. The direction of hand motion (Kalaska *et al.*, 1989, Georgopoulos, 1988, Kettner *et al.*, 1988, Georgopoulos *et al.*, 1988, Georgopoulos *et al.*, 1983, Georgopoulos *et al.*, 1982), the static hand position (Kettner *et al.*, 1988), and the direction of hand load resistance (Kalaska *et al.*, 1989) can be predicted by a "population vector" formed from a linear combination of the cortical cells' firing rates weighted by their preferred directions.

4. It is difficult to predict a cell's firing rate from the EMG response produced by microstimulation nearby (Murphy *et al.*, 1982).

Kettner *et al.* (1988) suggest that "the arm area of the motor cortex is mostly concerned with the planning and implementation of the direction of reaching in space." This claim is based on "observations that the activity of most single cells in that area relates in an orderly fashion to the direction of movement in space" (Kettner *et al.*, 1988, Schwartz *et al.*, 1988, Georgopoulos *et al.*, 1982).

At this point, I will restate in my notation some of the assumptions and results of much of this work (see (Schwartz *et al.*, 1988, Georgopoulos *et al.*, 1988, Kettner *et al.*, 1988)). Some assumptions are:

A1. Cell firing rate χ_i is a smooth function of desired hand direction m_d for this task:

$$\chi_i = \chi_i(m_d)$$

where $m_d \in R^3$ is the unit vector in the direction of desired motion.

A2. There is a distribution of peaks of direction sensitivity:

$$\xi_i = \arg\max_{m_d}(\chi_i(m_d)).$$

A3. Each cell, when stimulated, tends to move the hand in a particular direction m_h:

$$c_i = m_h(\chi_i)$$

where $m_h \in R^3$ is the unit vector in the direction of actual hand motion.

A4. The actual hand direction m_h is the same as the desired hand direction m_d:

$$m_h = m_d.$$

A5. The distributed representation is a unique function of the actual hand direction m_h independent of the current task:

$$\chi = \chi(m_h).$$

Assumptions A1 and A4 imply that the motor cortex embeds low-dimensional information about hand direction in a high dimensional space of many neurons. This means that the motor cortex is assumed to use a distributed representation of hand direction. (This is a direct consequence of the fact that multiple cortical cells will respond for hand motion in any given direction.) Assumption A2 implies that the basis functions in the high dimensional space are local. Assumptions A1 through A3 state that the coordinates of both the input and output spaces have been chosen by the experimenters to be the direction of hand motion. Assumption A4 implies that not only does the animal correctly solve the task from its own point of view, but that its internal representation of the task must be in spatial coordinates.

The results are:

R1. Cell firing can be approximated by a cosine function of the desired hand position:

$$\chi_i \approx a_i + b_i \cos(\text{angle}(m_d) - \text{angle}(\xi_i))$$

(where the subtraction is performed on the sphere, and a_i and b_i are constants).

R2. There is an approximately uniform distribution of centers ξ_i.

R3. The actual hand motion can be approximated by a linear combination of the cell firing rates:

$$m_h \approx \sum_i c_i \chi_i = C\chi$$

where the coefficients are equal to the preferred direction of each cell:

$$c_i = \xi_i$$

giving:

$$m_h \approx \sum_i \xi_i \chi_i.$$

R4. The cell firing rates cannot be predicted from the EMG data

$$\chi_i \neq \chi_i(\text{EMG}).$$

Given assumptions A1 through A3, R1 implies that the basis functions are in fact radial basis functions, since cosine can be thought of as a local function (on the sphere of possible directions). From this we know that it must be possible to approximate any sufficiently smooth function on the sphere as a linear combination of the radial basis functions, so without further experimentation we can conclude that

$$m_h \approx \sum_i c_i \chi_i$$

for some set of vectors c_i. If we now add assumption A4 and result R2, then (Georgopoulos *et al.*, 1988, Mussa-Ivaldi, 1988) proved that these are sufficient conditions for

$$m_h \approx \sum_i \xi_i \chi_i$$

so that we can conclude $c_i = \xi_i$. Therefore, result R3 is a theoretical consequence of results R1 and R2, given the assumptions. This is a main result of this paper. I have shown that the fact that the population vector predicts hand motion is a necessary result of the assumptions, given R1 and R2.

4.2   THE INTERNAL REPRESENTATION

I will now examine some of the consequences of assumption A5. If we assume that the hand motion can somehow be related to EMG data, so that

$$m_h = m_h(\text{EMG})$$

then result R4 suggests that assumption A5 is incorrect in this form. What assumption A5 basically means is that the cell firing rates should always be determined by an internal representation of the desired trajectory in external coordinates. For tasks such as reaching to a fixed target in space, this is perhaps a useful representation. But it is important to realize that there will be other tasks for which this is not necessarily a good representation. There is evidence that motor cortical cells respond well to a variety of sensory modalities, including visual (Wannier *et al.*, 1989), proprioceptive (Flament and Hore, 1988), and tactile (Huang *et al.*, 1989). As discussed in section 3, there can be multiple different representations which lead to the same output, and these representations can be chosen to make any particular task easier.

These considerations help to eliminate a possible inconsistency in different researchers' results. (Schwartz *et al.*, 1988) and (Georgopoulos *et al.*, 1988) found a linear relationship between direction of movement and cortical cell firing. (Kettner *et al.*, 1988) found a linear relationship between absolute position in space and cell firing. (Kalaska *et al.*, 1989) found a linear relationship between load direction and cell firing. And (Flament and Hore, 1988) found a relationship between arm acceleration and cell firing.

There are many possible explanations for this apparent inconsistency. One, suggested by (Kettner *et al.*, 1988), is that the tonic and phasic components of cell firing encode different properties of the intended movement. Another possibility is that the actual output of the cortex controls an equilibrium point for the hand position, and so hand static position, motion direction, and force direction are all equivalent (Bizzi *et al.*, 1986, Flash, 1987, Whitney, 1987). Yet another possibility suggested in (Mussa-Ivaldi, 1988) is that in fact the controlled variable may be change in muscle length, which is linearly related to the direction of hand motion.

I would like to add a final possibility that the distributed representation implies that there may not be a unique representation in the cortex, and that many possible internal representations can lead to the same motor behavior (see section 3). In this case, the internal representation may depend on the task presented to the animal. The best control strategy may differ for different tasks. Sometimes this control strategy may need to make use of visual, tactile, or proprioceptive information, while at other times this may not be necessary. Microelectrode recordings of cells will reveal the current control strategy, but do not necessarily generalize beyond the current task.

Since distributed representations allow linear approximation of arbitrary (smooth) functions, it is no surprise that the direction of motion, load, or static position can be computed as a linear function of the cells' firing rates. However, the actual computation ρ which links motor cortical cells to motor units in muscles may be quite complex. It is known that any given corticospinal axon may terminate on alpha motoneurons of several different muscles (Shinoda *et al.*, 1981). So activation of any single cortical cell may not lead to particularly useful behavior (Georgopoulos *et al.*, 1988). The activity of an entire population of cells controlled according to a particular strategy will be necessary to achieve useful motion.

4.3 Interpretation of Results

So what can we learn from the experimental results? The two most significant results are that the direction specificity of neurons is distributed approximately uniformly on the sphere, and that the tuning for any particular neuron looks like a raised cosine function. The importance of these two conclusions comes from the fact that neither one is necessarily required for computational purposes. Even with a highly non-uniform distribution Φ (even one which did not satisfy the sufficient requirements given in (Georgopoulos *et al.*, 1988) and (Mussa-Ivaldi, 1988)), $\rho(\Phi(x))$ may still approximate $f(x)$. Similarly, the tuning function might not have to be a cosine function, but perhaps could instead be any radially symmetric function of the difference in movement direction. It is possible that this form of $\Phi(x)$ is required by the output function ρ and the desired function $f(x)$, but the particular symmetries of this representation suggest that we should look for other biological constraints on the system which might lead to these results.

One possible explanation for the uniform distribution of movement directions is based on symmetry arguments. For tasks near the center of an animal's workspace, in which presumably all directions are useful, there is no good way to distinguish any particular set of directions from any other. In this case, we might expect that the neural representation would be symmetric with respect to rotations of the hand coordinate frame. This would lead naturally to a uniform distribution. In regions of the workspace which approached workspace limits (such as joint limits) or in which the animal was not accustomed to working (such as behind the head) we might expect that there could be anisotropies in the distribution of preferred directions. Any control strategy for operating in these regions would have to take this into account in order to generate appropriate hand trajectories.

Another possible explanation might be that a uniform distribution is useful for learning. A uniform distribution implies that hand coordinates are a linear function of cell firing rates, where the coefficients are equal to the preferred directions for any particular cell (Mussa-Ivaldi, 1988). (Although hand direction will always be a linear function of firing rates, for non-uniform distributions the coefficients are not necessarily related to the preferred directions given by the control strategy. (Kalaska *et al.*, 1989) figure 10B shows an example where the population vector computed for a non-uniform distribution is inaccurate.) If a learning task is specified in terms of hand coordinates, then in order to learn a control strategy for that task the animal must invert the relationship between trajectories and a surface in the distributed representation. The fact that the preferred directions for the input map and the coefficients of the output map are the same may allow a great simplification of learning strategies, since each map can be derived from the other.

I now turn to the raised cosine tuning curves for each cell. Note that a raised cosine is not a local basis function in R^n, but that it is local if we consider it to represent the response on the sphere as a function of angle between two vectors. The raised cosine function is the smoothest possible function in the sense of section 2.3. As mentioned there, a basis function of this type does not allow approximation of functions with any higher harmonics. Another way to see this is to realize that for any two unit vectors u and v, the dot product $u^T v$ is equal to the cosine of the angle between the two vectors. So we can write the experimentally approximated cosine function

$$h(\|m_d - \xi_i\|) = a_i + b_i \cos(\text{angle}(m_d) - \text{angle}(\xi_i))$$

as

$$h(\|m_d - \xi_i\|) = a_i + b_i \xi_i^T m_d$$

so long as we restrict ourselves to having ξ_i and m_d be unit vectors. In this case we see that the function $\Phi(x)$ is linear (affine, actually), so the set of functions which we can compute with a linear output map C is restricted to be the set of linear functions. If the set of possible inputs lies on the surface of a sphere (since $\|m_d\| = 1$), then the set of possible outputs will lie on the surface of an ellipsoid. Since data was only gathered for samples of m_d with unit length (ie: all movements had the same distance and presumably the same average velocity), there is no way to know if the linear approximation to the cell firing rate is valid anywhere outside the surface of the sphere $\|m_d\| = 1$.

Even so, for this restricted input set this particular choice of control strategy means that for arbitrary $f(x)$, it is unlikely that f can be approximated by $C\Phi(x)$ for linear C. The fact that we can make this approximation implies that $\rho(\Phi)$ has a special form for outputs in hand coordinates. It appears that ρ itself may be a linear function within the subspace of movement directions and velocities tested.

A numerical explanation for the cosine tuning can be derived from Taylor series expansions. For small perturbations, any nonlinear function $\Phi(x)$ can be approximated by the linear term of its Taylor expansion $J(x)x$ where $J(x)$ is the Jacobian matrix $d\Phi/dx$ of Φ evaluated at x. In other words, so long as the target remains in a small region of space, the mapping $\Phi(x)$ will appear to be linear, and so the tuning curves will be cosines. For large deviations of the target, $\Phi(x)$ will no longer be linear, and the tuning curves (if they are tuned at all) will no longer be cosines. In fact, recent data from (Caminiti, 1990) indicate that the tuning curves are approximated locally by cosines but that the preferred directions change as the targets move to different regions of the workspace. This corresponds to the idea that locally $J(x)$ may be constant, but that for large changes in x the tuning $J(x)$ will change (while still appearing locally linear).

5   DISCUSSION

The theory in sections 2 and 3 allows us to make three important predictions about distributed representations. First, if the cortex uses population coding, then we expect that it will be possible to approximate sufficiently smooth functions as linear combinations of the response of cells in the distributed representation. Second, there are many different possible control strategies which will achieve a given motor performance, and the choice of control strategy, and hence the firing rate of individual cells, may depend on the task which an animal is performing. Third, the computational power of distributed representations depends on the nonlinear properties of these representations, so we do not expect a distributed representation to be a linear function of the input unless there are additional constraints on the system which require this.

Now, the results stated in section 4 allow us to make three more specific statements about the representation of movements in motor cortex. First, motor cortex appears to use a representation in which low-dimensional variables are population coded in a high-dimensional space. Second, the distribution of directions to which cells in the representation respond maximally is approximately uniform. And third, the tuning of many cells can be described locally as a raised cosine of the difference between the desired hand direction and a preferred direction for each cell.

When we apply the theory to the data, we can draw the following conclusions. First, the representation which we discover is not unique, since it depends on both the task and the choice of input representation. Second, the output mapping is not unique since it depends on the output representation and the control strategy. Third, the fact that the output map is linear does not give us much information, since we expect that it will be approximately linear for output representations which are sufficiently smooth functions of the input representation. Fourth, the uniform distribution of preferred directions and the raised cosine tuning are not predicted by theoretical considerations, and thus lead us to look for biological constraints to explain these two phenomena.

In designing experiments to look at distributed representations, we see that it is very important to specify that any measurement of cell firing must be related to the task being performed. The structure of distributed representations implies that the choice of the input and output coordinates will actually determine the results obtained, as will the control strategy being used. A further consideration is to realize that the input map Φ and the output map ρ which controls the muscles are different, and require different investigational techniques. Recording studies give information only about the control strategy defined by the input map. Microstimulation studies are necessary to determine uniquely the control function of any given cell or group of cells in motor cortex.

As a computational technique, population coding allows considerable flexibility in approximating functions using linear maps. So long as all transformations are smooth, it may be possible to approximate arbitrary output functions, make arbitrary changes of coordinates, or combine information from multiple different coordinate systems (modalities). The ability to interconvert coordinate systems can be very useful for any computational system (Knudsen et al., 1987). It is important to keep in mind the restriction of "sufficiently smooth", however. As we saw above, if the basis functions are too smooth (cosine functions, for example), then the set of functions which can be linearly computed may be quite restricted. In this case, the output map ρ will not be approximated by a linear function, and arbitrary linear changes of coordinate will be impossible. If it turns out that ρ can still be approximated linearly, then this gives a clue to the structure of the nonlinear map ρ itself.

My discussion here suggests certain experiments which may be useful for understanding how the motor cortex processes information. It would be interesting to investigate what different representations exist in motor cortex, and how exactly these representations depend upon different tasks. This would mean recording from cells during behavior of different activities. Another experiment would be to determine if different control strategies could be used to solve the same task, if different task specifications are given. In the studies presented here, the sensory information (position of a light) was always in the same coordinate system as the motor response (move the hand to a target at the light). These two variables should be separated, perhaps by having the desired target specified in terms of a written symbol, or even a tactile stimulus. It appears that motor cortical activity occurs during the planning stages of a movement (Georgopoulos et al., 1989b, Georgopoulos et al., 1989a), and experiments of this nature might indicate whether this planning occurred in terms of stimulus coordinates or response coordinates. Finally, it is important to investigate situations in which the theory presented here may not apply. It would be useful to know whether there exist tasks or variables which are not population coded. And it would be important to see if there are useful output variables which are not "sufficiently smooth" and thus are not linear functions of the distributed representation.

REFERENCES

Andersen R. A., Zipser D., 1988, The role of the posterior parietal cortex in coordinate transformations for Visual-Motor integration, *Can. J. Physiol. Pharmacol.*, 66:488–501.

Andersen R. A., Essick G. K., Siegel R. M., 1985, Encoding of spatial location by posterior parietal neurons, *Science*, 230:456–458.

Bizzi E., Mussa-Ivalda F. A., Hogan N., 1986, Regulation of multi-joint arm posture and movement, *Prog. Brain Res.*, 64:345–351.

Broomhead D. S., Lowe D., 1988, Multivariable functional interpolation and adaptive networks, *Complex Systems*, 2:321–355.

Caminiti R., 1990, Making arm movements within different parts of space: Dynamic mechanisms in the primate motor cortex, manuscript in preparation.

Farmer J. D., Sidorowich J. J., 1989, Predicting chaotic dynamics, In Kelso J. A. S., Mandell A. J., Shlesinger M. F., ed.s, *Dynamic Patterns in Complex Systems*, pages 265–292, World Scientific.

Flament D., Hore J., 1988, Relations of motor cortex neural discharge to kinematics of passive and active elbos movements in the monkey, *J. Neurophys.*, 60(4):1268–1284.

Flash T., 1987, The control of hand equilibrium trajectories in multi-joint arm movements, *Biol. Cyb.*, 57:257–274.

Georgopoulos A. P., Kalaska J. F., Caminiti R., Massey J. T., 1982, On the relations between the direction of Two-Dimensional arm movements and cell discharge in primate motor cortex, *J. Neurosci.*, 2(11):1527–1537.

Georgopoulos A. P., Caminiti R., Kalaska J. F., Massey J. T., 1983, Spatial coding of movement: A hypothesis concerning the coding of movement direction by motor cortical populations, *Exp. Brain Res. Suppl. 7*, pages 328–336.

Georgopoulos A. P., Kalaska J. F., Caminiti R., 1985, Relations between Two-Dimensional arm movements and Single-Cell discharge in motor cortex and area 5: Movement direction versus movement end point, *Exp. Brain Res. Suppl. 10*, pages 175–183.

Georgopoulos A. P., Kettner R. E., Schwartz A. B., 1988, Primate motor cortex and free arm movements to visual targets in Three-Dimensional space. II. coding of the direction of movement by a neuronal population, *J. Neurosci*, 8(8):2928–2937.

Georgopoulos A. P., Crutcher M. D., Schwartz A. B., 1989a, Cognitive Spatial-Motor processes, *Exp. Brain Res.*, 75:183–194.

Georgopoulos A. P., Lurito J. T., Petrides M., Schwartz A. B., Massey J. T., 1989b, Mental rotation of the neuronal population vector, *Science*, 243:234–236.

Georgopoulos A. P., 1986, On reaching, *Ann. Rev. Neurosci.*, 9:147–170.

Georgopoulos A. P., 1988, Neural integration of movement: Role of motor cortex in reaching, *FASEB J.*, 2:2849–2857.

Grossberg S., 1976, On the development of feature detectors in the visual cortex with applications to learning and reaction-diffusion systems, *Biological Cybernetics*, 21:145–159.

Hinton G. E., McLelland J. L., Rumelhart D. E., 1986, Distributed representations, In McLelland J. L., Rumelhart D. E., The PDP Research Group , ed.s, *Parallel Distributed Processing*, pages 77–109, MIT Press, Cambridge, MA.

Huang C. S., Hiraba H., Sessle B. J., 1989, Input-output relationships of the primary face motor cortex in the monkey, *J. Neurophys.*, 61(2):350–362.

Kalaska J. F., Caminiti R., Georgopoulos A. P., 1983, Cortical mechanisms related to the direction of Two-Dimensional arm movements: Relations in parietal area 5 and comparison with motor cortex, *Exp. Brain Res.*, 51:247–260.

Kalaska J. F Cohen D. A. D., Hyde M. L., Prud'homme M., 1989, A comparison of movement direction-Related versus load Direction-Related activity in primate motor cortex, using a Two-Dimensional reaching task, *J. Neurosci.*, 9(6):2080–2102.

Kalaska J. F., 1988, The representation of arm movements in postcentral and parietal cortex, *Can. J. Physiol. Pharmacol*, 66:455–463.

Kawato M., Furukawa K., Suzuki R., 1987, A hierarchical neural-network model for control and learning of voluntary movement, *Biol. Cyb.*, 57:169–185.

Kettner R. E., Schwartz A. B., Georgopoulos A. P., 1988, Primate motor cortex and free arm movements to visual targets in Three-Dimensional space. III. positional gradients and population coding of movement direction from various movement origins, *J. Neurosci.*, 8(8):2938–2947.

Klopfenstein R. W., Sverdlove R., 1983, Approximation by uniformly spaced gaussian functions, In Chui C. K., Schumaker L. L., Ward J. D., ed.s, *Approximation Theory IV*, pages 575–580, Academic Press.

Knudsen E. I., du Lac S., Esterly S. D., 1987, Computational maps in the brain, *Ann. Rev. Neurosci.*, 10:41–65.

Kohonen T., 1982, Self-organized formation of topologically correct feature maps, *Biological Cybernetics*, 43:59–69.

Kohonen T., 1988, The "neural" phonetic typewriter, *Computer*, 21:11–22.

Lacquaniti F., 1989, Central representations of human limb movement as revealed by studies of drawing and handwriting, *TINS*, 12(8):297–281.

Linsker R., 1989, How to generate ordered maps by maximizing the mutual information between input and output signals, *Neural Computation*, 1:402–411.

McIlwain J. T., 1975, Visual receptive fields and their images in superior colliculus of the cat, *J. Neurophys.*, 38:219–230.

Miller W. T., Glanz F. H., Kraft L. G., 1987, Application of a general learning algorithm to the control of robotic manipulators, *Intl. J. Robotics Res.*, 6(2):84–98.

Moody J., Darken C., 1988, Learning with localized receptive fields, Technical Report RR-649, Yale Dept. Computer Science.

Moody J., Darken C., 1989, Fast learning in networks of Locally-Tuned processing units, *Neural Computation*, 1:281–294.

Moody J., 1989, Fast learning in Multi-Resolution hierarchies, Technical Report RR-681, Yale U.

Murphy J. T., Kwan H. C., MacKay W. A., Wong Y. C., 1982, Precentral unit activity correlated with angular components of a compound arm movement, *Brain Res.*, 246:141–145.

Mussa-Ivaldi F. A., 1988, Do neurons in the motor cortex encode movement direction? an alternative hypothesis, *Neurosci. Lett.*, 91:106–111.

Penrose R., 1955, A generalized inverse for matrices, *Proc. Cambridge Philos. Soc.*, 51:406–413.

Poggio T., Girosi F., 1990, Regularization algorithms for learning that are equivalent to multilayer networks, *Science*, 247:978–982.

Powell M. J. D., 1987, Radial basis functions for multivariable interpolation: A review, In Mason J. C., Cox M. G., ed.s, *Algorithms for Approximation*, pages 143–167, Clarendon Press, Oxford.

Raibert M. H., 1978, A model for sensorimotor control and learning, *Biol. Cyb.*, 29:29–36.

Renals S., Rohwer R., 1989a, Learning phoneme recognition using neural networks, Proc. ICASSP-89.

Renals S., Rohwer R., 1989b, Phoneme classification experiments using radial basis functions, Proc. IJCNN-89.

Ritter H., Schulten K., 1988a, Convergence properties of kohonen's topology conserving maps: Fluctuations, stability, and dimension selection, *Biol. Cyb.*, 60:59–71.

Ritter H., Schulten K., 1988b, Extending kohonen's Self-Organizing mapping algorithm to learn ballistic movements, In Eckmiller R., v. d. Malsburg C., ed.s, *Neural Computers*, pages 393–406, Springer-Verlag, Berlin.

Ritter H. J., Martinetz T. M., Schulten K. J., 1989, Topology-conserving maps for learning visuo-motor-coordination, *Neural Networks*, 2:159–168.

Sanger T. D., 1990a, Learning nonlinear features using eigenvectors of radial basis functions, submitted to *IEEE Trans. Neural Networks*.

Sanger T. D., 1990b, Theoretical considerations for the analysis of population coding in motor cortex, submitted to *Neuroscience Letters*.

Saund E., 1989, Dimensionality-reduction using connectionist networks, *IEEE PAMI*, 11(3):304–314.

Schwartz A. B., Kettner R. E., Georgopoulos A. P., 1988, Primate motor cortex and free arm movements to visual targets in Three-Dimensional space. I. relations between single cell discharge and direction of movement, *J. Neurosci*, 8(8):2913–2927.

Shinoda Y., Yokota J.-I., Futami T., 1981, Divergent projection of individual corticospinal axons to motoneurons of multiple muscles in the monkey, *Neurosci. Lett.*, 23:7–12.

Steinmetz M. A., Motter B. C., Duffy C. J., Mountcastle V. B., 1987, Functional properties of parietal visual neurons: Radial organization of directionalities within the visual field, *J. Neurosci.*, 7(1):177–191.

van Gisbergen J. A. M., van Opstal A. J., Tax A. A. M., 1987, Collicular ensemble coding of saccades based on vector summation, *Neuroscience*, 21(2):541–555.

Wannier T. M. J., Maier M. A., Hepp-Reymond M.-C., 1989, Responses of motor cortex neurons to visual stimulation in the alert monkey, *Neurosci. Lett.*, 98:63–68.

Whitney D. E., 1987, Neuronal coding and robotics, *Science*, 237:300–301.

ACKNOWLEDGEMENTS

Thanks are due to Sandro Mussa-Ivaldi, Emilio Bizzi, Richard Lippmann, and Marc Raibert for their suggestions and comments. This report describes research done within the laboratory of Dr. Emilio Bizzi in the department of Brain and Cognitive Sciences at MIT. The author was supported during this work by the division of Health Sciences and Technology, and by NIH grants 5R37AR26710 and 5R01NS09343 to Dr. Bizzi.

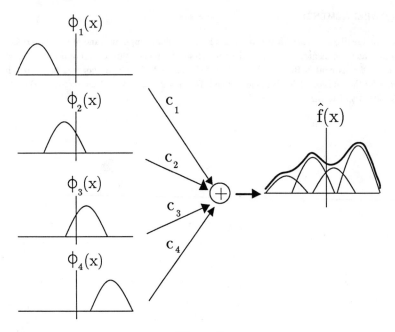

Figure 1

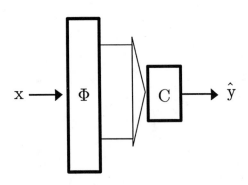

Figure 2

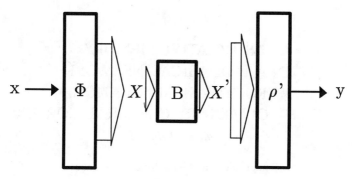

Figure 3

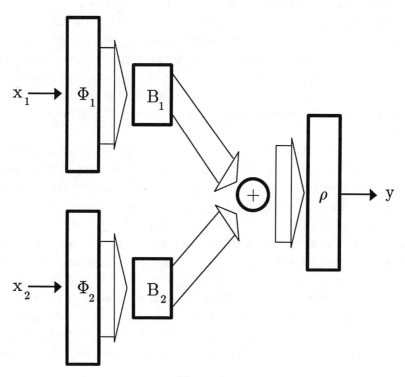

Figure 4

81

4

ASSOCIATIVE MEMORY IN A BIOLOGICAL NETWORK: STRUCTURAL SIMULATIONS OF THE OLFACTORY CEREBRAL CORTEX

James M. Bower

Computation and Neural Systems Program
California Institute of Technology
Pasadena, CA. 91125

1. Olfactory Object Recognition and Associative Memory

One of the primary objectives of research in our laboratory is to understand how memory function is implemented in real biological networks. We have chosen to pursue this question by exploring the ability of the mammalian olfactory system to recognize objects based on the complex blends of airborne molecules they emit. This process is believed to involve the construction of an associative memory that uses previous olfactory experience to recognize current olfactory stimuli (Haberly, 1985; Haberly and Bower 1989).

The view that the olfactory system depends on an associative memory process is based on both a general consideration of the computational task faced by this system as well as on the overall structure of its neural circuits. At the most abstract level, olfactory object recognition involves identifying the chemically diverse blends of airborne molecules emitted by different objects (Lancet, 1986; Laing et al., 1989). While very little is yet known about the natural structure of "olfactory stimulus space", olfactory discrimination certainly requires that the olfactory system be capable of recognizing diverse subsets of a large number of airborne molecules (Lancet, 1986; Laing et al., 1989).

One possible way to recognize different combinations of chemicals would be to establish a system of highly specific chemical "feature detectors" and hardwire the comparisons between them. Such an approach has actually been demonstrated in biological olfactory systems under certain special circumstances. However, the use of this strategy

This work was supported by ONR contract N00014-88-K-0513. The author acknowledges Matthew Wilson who is principally responsible for the modeling work on which this chapter is based and thanks Matt, Mark Nelson, and Michael Hasselmo for their critical reading of this manuscript.

appears to be restricted to the detection of blends of pheromones generated for communication between animals of the same species (Christensen and Hildebrand, 1987; Epple et al., 1989). In this case, however, the sender of the signal is motivated to limit the variability in the chemical signal, which greatly simplifies the task of the receiver.

A very different situation appears to apply in the case of more general odor detection. In this instance, the olfactory system has to cope with a considerable amount of variability as well as a tremendous diversity of chemical combinations that could be biologically relevant (Lancet, 1986; Laing et al., 1989). Thus, in the case of general olfaction, it seems unlikely that a system based on detecting a subset of specific odorants would be very successful. Instead, it seems more appropriate to develop a computational strategy which relies on broadly sampling the olfactory stimulus space and then learning associations between the different chemical signals emitted by particular objects. Associative memory networks would be a quite an appropriate way to implement this computational strategy.

The structure of the neural networks that constitute the general purpose olfactory system do, in fact, suggest that something like an associative network is being implemented. In particular, the piriform cortex, which as the primary olfactory cerebral cortex has been a major focus of the research effort in our laboratory, seems very well suited to implementing an associative memory (Tanabe et al., 1975; Haberly 1985; Haberly and Bower, 1989; Bower, 1990). As described in more detail below, its structure supports an extensive divergence and convergence of neuronal processes which could very well serve as a substrate for associating activity evoked by many different combinations of chemically diverse molecules. In this way, the piriform cortex also very generally resembles numerous abstract "neural net" models that have been specifically constructed to implement associative memories (Palm, 1980; Hopfield,1982; Kohonen, 1984; Grossberg, 1988). The next section will consider the structure of piriform cortex in this context.

2. General Structure of Piriform Cortex and the Olfactory System

In order to substantiate any specific link between the piriform cortex and associative memory function it will be necessary to explore how well the detailed structure of this network might support such a function. This effort is the main focus of current work in our laboratory and is the central subject of this chapter. The following section considers briefly the relationship between piriform cortex and other brain regions as well as describing very generally the structure of its circuits. Readers requiring more than the cursory anatomical description presented here are referred to the review by Haberly (1985).

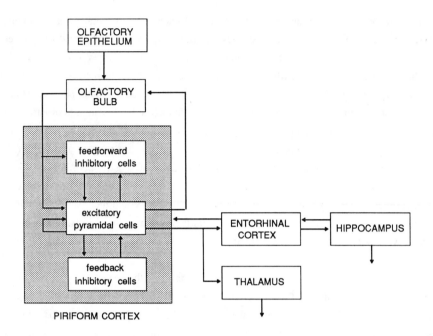

Figure 1 This schematic diagram demonstrates several of the pathways providing input to and output from the piriform cortex. The gray shaded box also indicates the principle cellular components of the model of piriform cortex discussed in the chapter. (Figure reprinted with permission from Bower 1990).

The first thing to note about the piriform cortex is that it occupies an unusual position among all primary cerebral cortical sensory areas in that it is positioned close to both the sensory periphery and deeply buried forebrain structures, like the hippocampus essential for more general memory function (Figure 1). With respect to its peripheral connections, information from the olfactory receptor neurons in the nasal epithelium is processed by only one intervening structure, the olfactory bulb, before it is sent to the piriform cortex. In all other sensory systems, peripheral sensory information reaches primary cerebral cortical areas only through the thalamus (Price, 1985). With respect to its outputs, a primary projection of piriform cortex activity is through the multisensory entorhinal cortex directly to the hippocampus (Luskin and Price, 1983; Price, 1985; Insausti et al., 1987). The hippocampus, which is believed to be crucial for general properties of memory storage in the nervous system (Squire, 1986a; 1986b; McNaughton and Morris, 1987), is influenced by other sensory modalities through entorhinal cortex, but only after many more processing stages. Thus the olfactory system effectively represents a short cut from the sensory periphery to deeply buried forebrain structures. This may provide an explanation for the well known ability of olfactory stimuli to evoke very strong memories (Cain, 1976).

The general purpose olfactory system is also unusual as a sensory system in the overall response properties of its neurons. Beginning at the sensory periphery, individual olfactory receptors respond to a remarkably broad range of chemical stimuli (Gesteland et al.,1965; Sicard and Holley; 1984; Lancet, 1986; Duchamp - Viret, 1989). Individual neurons in both the olfactory bulb and piriform cortex are also remarkably broadly tuned in their responses to different chemical stimuli (Tanabe et al., 1975) giving the impression, again, that the general purpose olfactory system is not based on a well organized population of feature detectors.

The apparent intermixing of responses to different peripheral stimuli at the level of single neurons is also reflected in the overall spatial organization of population responses within olfactory networks. For example, there is very little evidence in olfaction for the detailed patterns of spatial organization seen in response to peripheral stimulation within early processing stages of the visual or somatosensory systems (Mountcastle, 1957; Hubel and Wiesel, 1968. While there is still some debate as to the existence of a very broad spatial pattern of stimulus feature mapping in the olfactory bulb (Jastreboff et al., 1984; Kauer, 1987), it is generally accepted that there is no such organization in the piriform cortex (Scott et al., 1980; Haberly and Bower, 1989).

The broad overlapping patterns of afferent induced physiological activity in the olfactory system are also reflected in the anatomical organization of its circuits. This is especially true in the case of the piriform cortex with respect to both the projections it receives from the olfactory bulb and the intrinsic excitatory connections within the cortical network itself (Devor, 1976; Haberly and Price, 1978; Luskin and Price, 1983). There is no apparent spatial organization in afferent projections from the olfactory bulb to the cortex. All evidence indicates that any location in the olfactory bulb is just as likely to innervate neurons in any region of the cortex. Similarly, the intrinsic excitatory pyramidal cell to pyramidal cell connections within the cortex itself are also highly diffuse and overlapping. In both respects, again, the piriform cortex is quite different from other primary sensory cerebral cortical areas (Gilbert, 1983). This unusual divergence and convergence of information must clearly be taken into account in trying to understand the function of these networks.

3. Computer Simulations of the Olfactory Cortex

The central feature of our effort to quantify the possible relationship between the structure of the olfactory system and its presumed function involves the construction of realistic computer simulations of piriform cortex (Bhalla et al., 1988; Bower et al., 1988; Wilson and Bower, 1988; 1989; Nelson et al., 1989; Bower, 1990). These simulations are based on what is currently known of the actual anatomical and physiological features of this network (Wilson and Bower, 1989). As such, the models serve as a means to explore the functional consequences of known network structure (Wilson and Bower, 1988; 1989) and also serve to identify the additional experimental information that is necessary to advance our understanding of the system (Hasselmo and Bower, 1990). In this way, the models suggest interpretations of current results as well as providing direction for future experimental work. Examples of both types of interactions are presented below.

Physically, the models built in the lab are computer-based compartmental numerical simulations. The piriform cortex model whose results are considered here, consists of the three principal types of neurons found in this cortex; the excitatory pyramidal cell and two different types of inhibitory neurons (Figure 2). The physiological properties of these neurons and their synaptic connections are taken from real physiological data, as are the geometrical relationships between the different neurons in the cortex. In particular, the model duplicates the extensive and broadly distributed bulbar projections to the cortex as well as the similarly arranged intrinsic excitatory connections within the network itself (Figure 3). A typical model run consists of up to several thousands of each type of neuron. Model output is in the form of physiologically measurable signals like neuronal spike trains, EEGs, and evoked potentials. Readers interested in details of the mathematical structure of this model are referred to Wilson and Bower (1989).

In the early stages of this modeling effort, the principal objective was to replicate the general physiological response properties of piriform cortex under a variety of stimulus conditions (Wilson and Bower, 1990). This work, which will not be described here, involved replicating the periodic or oscillatory behavior induced in the cortex by both direct electrical stimulation of its afferents and natural activation of its inputs (Freeman, 1975). Besides suggesting new mechanisms for the generation of this periodic activity, this initial modeling effort also increased confidence that the simulations captured essential structural and physiological cortical features (Wilson and Bower, 1990). This, in turn, has led to preliminary investigations into the possible functional significance of the details of cortical structure for olfactory object recognition (Wilson and Bower, 1988).

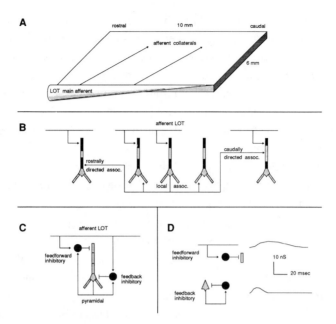

Figure 2 This figure schematically represents the structural features of piriform cortex which have served as the basis for the simulations discussed in the text. The diagram at the top indicates the spatial pattern of projection of lateral olfactory tract (LOT) axons into and across the cortex. The middle diagram shows the excitatory connections made by afferent and pyramidal cell axons within the cortex and simulation. Note that pyramidal cell "association" connections are both local and distant. The schematic at the bottom left indicates the basic pattern of interconnections between pyramidal cells and the two classes of inhibitory neurons modeled. The diagram at the bottom right shows the different temporal properties of the synaptic conductances induced by the two classes of inhibitory neurons. (Figure reprinted with permission from Bower 1990).

4. Modeling Piriform Cortex as an Associative Memory

As mentioned in the introduction, the extensive and spatially distributed pattern of afferent projections to olfactory cortex and the extensive distributed excitatory connections found within this cortical network are fully compatible with an associative memory function (Haberly and Bower, 1989). More abstract models designed over the last twenty years to explore associative memory function are usually based on similarly extensive intrinsic excitatory connections (Palm 1980; Hopfield, 1983; Kohonen, 1984). In fact, these more abstract models not only share a certain general structural similarity with the piriform cortex, they also share certain computational capabilities with olfactory processing. For example, abstract associative memory models can be made to recognize objects in the presence of relatively high levels of background noise just as the olfactory system is capable of operating in extremely noisy environments. Abstract associative models have also been shown to be capable of reconstructing original learned patterns from inputs that are fragmented or distorted (Kohonen et al., 1976; Kohonen, 1984). As described below, this ability to do so-called "pattern completion" is probably also important to olfactory pattern recognition.

While the associative functioning of the olfactory system, just like the natural olfactory stimuli it interprets, is highly complex and varied, our initial investigations of the learning capacity of the piriform cortex model has concentrated on two relatively simple aspects of associative learning. First, the model's ability to generate consistent patterns of neuronal activity in response to specific input patterns has been investigated. Presumably, if the olfactory cortex is responsible for odor recognition, it should be able to generate consistent neuronal output in the presence of consistent sensory input. Second, the ability of the model to generate a stable pattern of neuronal activity in the presence of slight changes in the input stimulus was studied. Because the mix of molecules being emitted by any object can vary with, for example, its age or environmental circumstances, it is presumed that olfactory recognition must, to some extent, be insensitive to such variations. As mentioned above, abstract associative memory models are capable of both types of behavior. The question at the center of this work is how a network structured like the olfactory cortex might perform similar pattern recognition and pattern completion tasks.

To explore these questions, the olfactory cortex model was provided with input intended to loosely represent the activity of single neurons in the olfactory bulb. Synaptic connections from these putative bulbar neurons to neurons in the cortical model were assigned completely randomly as were the initial weights of each connection (see Figure 3). In order to explore learning in this network, a Hebb-type correlation learning rule was also introduced to govern activity dependent changes in the synaptic strengths of modeled connections (Hebb 1949). At the time when these simulations were performed, no information was yet available on the existence or form of synaptic modification in piriform cortex (Bower and Haberly, 1986), but evidence for Hebb-type synaptic modification did exist in the closely related hippocampus (Wigstrom et al., 1986). Additional information on the the actual experimental modeling conditions for these learning experiments are described in the legends for figures 3 and 4, and in Wilson and Bower (1988).

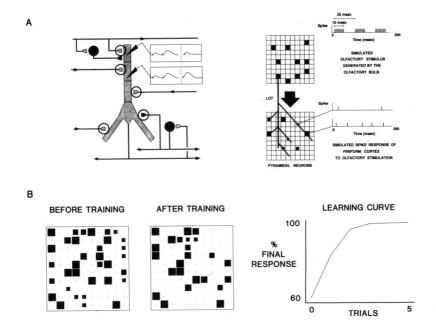

Figure 3. Results are shown from simulations in which the memory properties of the piriform cortex model were explicitly studied. **A** on the left indicates with circles those synapses undergoing activity dependent changes in connection strengths in these simulations (see text and Wilson and Bower, 1988). The simulated intracellular records attached to the diagramed electrodes indicate that the best performance of the network was achieved when long term Hebbian modifications were restricted to the association fiber system synapses. The diagram at the top right illustrates the random projection pattern of one of ten active bulbar neurons (100 total in the simulation) to cortical pyramidal cells is shown, and the stereotyped pattern of bulbar activity. The cortical section of this diagram demonstrates the display convention for neuronal activity used in figures 3 and 4. The size of each black box corresponds to the total number of action potentials generated by a cell in the corresponding position during 200 msecs of bulbar activity. **B** demonstrates the response of the simulated cortex to activity in a random set of 10 bulbar neurons. On the far left is shown the response of the cortex to the first presentation of the stimulus before training. The middle diagram indicates the final stable pattern of activity induced in the cortex following learning. The graph on the far right indicates the changes in the pattern of activity over the 5 trials necessary for this convergence to take place. (Figure reprinted with permission from Bower 1990.)

As mentioned above, the first objective of this modeling effort was to determine if the piriform cortex model was capable of learning to generate a stable output when presented with a consistent input pattern. Results of such an experiment are shown in figure 3. First, in the absence of synaptic modification, repeated presentation of the identical pattern of input activity resulted in a continually varying pattern of cortical activity (not shown). However, when a Hebbian-type (Hebb, 1949) synaptic modification was allowed under the same stimulus conditions, the network converged to a stable pattern of neuronal response after several stimulus presentations. As shown by the learning curve on the bottom right of figure 3, this convergence was relatively rapid, requiring only a small number of presentations. The top part of figure 4 demonstrates that the network could also learn to generate different patterns of activity in response to different patterns of input. While the total capacity of the network was not rigorously explored, it is encouraging that more than one pattern could be stored.

The second and perhaps more interesting question addressed with the model was to what extent a stable output, once learned, was resilient to changes in the input pattern. This was tested by reducing the number of active bulbar inputs by half and comparing the response under these conditions to the response of the network to the original full pattern of stimulation. General results are shown in the lower panels of figure 4. In these examples it can be seen that, before learning, reducing the number of active input neurons by 50% resulted in a pattern of activity in the cortex that was only 56% similar to the cortical response to a full stimulus. After learning, however, the same 50% reduction in active bulbar neurons produced a pattern of activity 80% similar to that evoked by the original input. Accordingly, the model demonstrates that once learned, specific cortical patterns can be regenerated when only a partial version of the original stimulus is presented. Again, natural variability in the molecules given off by an object presumably makes this capacity important in the recognition of olfactory stimuli.

5. Parameter variations and model predictions

With any model as complex as this model of piriform cortex, the question of parameter sensitivity becomes an important issue (Wilson and Bower, 1989). It must be pointed out, however, that many of the structural and physiological parameters used in the current model are constrained by the results of actual anatomical and physiological experiments. Taking advantage of these constraints, in fact, is one of the motivations for making models structurally realistic (Bower, 1990). However, as mentioned previously, one important feature of the model that was not initially constrained by actual data was the type and location of activity dependent synaptic learning within the network. At the time these simulations were conducted, it was not yet known which, if any of the different synapses in piriform cortex where capable of long term activity dependent changes in synaptic strength (Bower and Haberly, 1986). Accordingly, the model was used to compare the consequences of placing learning in different synaptic populations. The results demonstrate a distinct sensitivity to the particular conditions of synaptic learning that were assumed. In this way the model, in effect, generated a prediction for what should be found experimentally within the cortex.

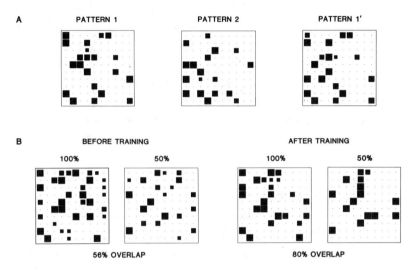

Figure 4. More results from the "memory" simulations described in figure 3 are shown. **A** demonstrates the simulated cortical response to successive training sessions using two different patterns of afferent input. The diagram at the far right demonstrates that the simulation retained its basic response to pattern 1 even after training with pattern 2. **B** shows the resilience of the simulation to degradation of the training input by 50% (i.e. only 5 of the original 10 bulbar neurons active). The diagrams on the left indicate the untrained initial response of the simulation to full and partial stimuli. The two diagrams on the right indicate responses to the same stimulus patterns after training. In this case training facilitated the ability of the cortex to generate a stable response pattern in the presence of significant degradation of the original input. Details of these results are discussed in Wilson and Bower 1988. (Figure reprinted with permission from Bower 1990.)

The specific objective of these modeling studies was to contrast the effects of incorporating a Hebb-type synaptic learning rule (Hebb, 1949) into either the afferent or association fiber system. As mentioned previously, these two fiber systems constitute the major excitatory synaptic influences within this cortex (Haberly and Bower, 1989) and they would presumably be essential components of any associative memory function (Haberly, 1985; Haberly and Bower, 1989). The results of these simulation experiments showed that, when synaptic modification was limited to synapses associated with the afferent fiber system, the ability of the network to generate a stable output pattern in response to a consistent input was greatly reduced as was the ability of the network to perform pattern completion (Wilson and Bower, 1988; Bower, 1990). When synapse modification was present in the association fiber system alone however, performance of both associative functions was greatly enhanced. Accordingly, the clear prediction from these results was that synapses of the intrinsic association fiber system should represent the principal site of synaptic learning in the olfactory cortex (Bower, 1990). Subsequent experimental results have indicated that this does in fact appear to be the case (Kanter and Haberly, 1989). Prompted by the accuracy of this prediction, additional physiological experiments in our laboratory, have revealed further differences in the physiological properties of these two types of excitatory synapses (Hasselmo and Bower, 1990). Preliminary simulations including these newly discovered synaptic differences suggest they also have important functional consequences. In this way, the interplay between simulations and experimental work has led to distinct progress in understanding this system.

6. Context Dependence in Olfactory Recognition

The previous discussion has centered around our attempts to model two relatively simple forms of associative memory function; the ability to generate a consistent output given an invariant input, and the ability to generate a stable output pattern in the presence of variations in the input pattern. While these two functions are presumably important to the process of olfactory recognition, interpreting chemical signals in the natural world is undoubtedly much more complex. For example, the resilience of the model to changes in the pattern of the peripheral input immediately raises the question as to how such a network would distinguish between objects of two different types emitting similar molecules.

In the jargon of computational vision, cases in which the data presented to the system are insufficient to unambiguously interpret the structure of the stimulus are referred to as being ill-posed problems (Marr and Hildreth, 1980). In the case of vision, one proposed resolution of such situations involves building into visual processing certain assumptions based on previous visual experience (Marr, 1982). For example, it has been proposed that the visual system may solve problems in interpreting the relative depths of different objects by assuming that objects are generally smooth with substantial discontinuities indicating borders between objects (Marr and Hildreth, 1980; Marr, 1982). In this way the visual system is presumed to make use of generally reliable relationships between objects in the visual scene. It seems unlikely, however, that such a strategy would necessarily be very useful in the olfactory system, where it is not clear that similar general rules exist concerning associations between different molecules in olfactory stimulus space. Rather, it seems more likely that the olfactory system evaluates each blend of odorants encountered as a unique event without relying on a more general interpretation based on its constituent parts. The fact that recognition events in olfaction are largely gestalt in nature, and that it is difficult to pick apart the subcomponents of a recognized olfactory signal supports this contention (Rabin and Cain, 1989; Richardson and Zucco, 1989).

Taken to its limit, the fundamental ambiguity of olfactory stimuli leads to the problem that in some or even many cases, the detailed molecular structure of the blend of odorants given off by a particular object may not contain enough information to support an accurate recognition of that object. If true, additional information would need to be obtained to solve the discrimination problem. We believe that this additional information is supplied by input from other sensory modalities which, in effect, establish a context in which to interpret olfactory information. Psychophysically, it is well known that olfactory recognition can be heavily influenced by the context in which the recognition takes place (Cain, 1976; Richardson and Zucco, 1989). It is also structurally characteristic of mammalian as well as insect olfactory networks that they project to and receive direct projections from regions of the brain in which other sensory modalities are highly corepresented. In mammalian systems, the entorhinal cortex, for example, is one of the most polymodal brain structures (Jones and Powell, 1970; Insausti et al., 1987).

Recently, we have used the piriform cortex model to begin to explore the possible role of contextual information in olfactory processing. To determine if contextual information could be used to make two structurally different input patterns generate a similar output, we paired each input during learning with a common "context" signal. In this case the contextual input was simply another pattern of 10 out of 100 active neurons taken to represent entorhinal projections. The results showed that following learning, network activity generated by either stimulus alone (i.e. without the contextual signal) was much more similar than it would have been had contextual information not been present during the learning phase of the simulation. Conversely, the model was also tested with two patterns of stimuli that were quite similar but were paired during learning with two different contextual patterns. In this case, subsequent stimulation with either training signal alone generated a pattern of cortical activity much less similar than it would have been had different contextual patterns not been given during training. Thus, the model does demonstrate the ability to modify the response of the network to afferent inputs in a way that is dependent on simultaneously presented contextual information. Analysis of the model indicates that the same features of the network responsible for pattern completion (Figure 4B) are also important in the modulation of cortical representations by contextual information (see Wilson and Bower, 1988). While these results represent only the first attempts to explore these more complex properties of olfactory object recognition, simulation results do reinforce the notion that context could be an important aspect of olfactory discrimination.

7. Comparison of Piriform Cortex With Other Primary Sensory Regions of Cerebral Cortex

In early sections of this chapter it was pointed out that the lack of stimulus related spatial organization in primary olfactory cortex was quite unusual when compared to other primary sensory cortical areas. In primary visual cortex, for example, there is a strict topographic organization of both afferent and intrinsic connections (Gilbert, 1983), and single neurons respond to highly restricted regions of visual space and to quite restricted stimulus properties (Hubel and Wiesel, 1968; 1972; Hubel, 1982). Presumably this reflects the importance of local computation in initial visual object recognition (Nelson and Bower, 1990). However, in later stages of visual processing, there appears to be a breakdown in the strict retinotopic organization of visual networks (Bruce et al., 1977; Gross et al., 1969; 1972; Desimone and Gross, 1976; 1979). In this way, these so-called "associative" regions of visual cortex begin to somewhat resemble the physical structure of piriform cortex. Further, neurons in these regions appear to prefer more complex visual stimuli that can be thought of as combinations of lower level stimulus features (Perrett et al., 1982; Hasselmo et al., 1989a; 1989b). Neurons in these regions also respond to these complex objects over larger extents of the visual field (Perrett et al., 1982; Desimone et al., 1984). In this way the visual system has been proposed to construct, through a series of processing stages, a progressively more complex representation of visual information (Marr, 1982). Output from the final stages of this visual processing then converges on brain regions like the hippocampus that are responsible for more general brain functions like memory.

In contrast to the visual system, the olfactory system, from the beginning, constructs complex representations of sensory data which then serve as a direct influence on deeply buried brain structures. Thus, piriform cortex appears almost to compress the role of multiple visual areas. Clearly, this difference in the two systems is related to fundamental differences in the computational problems faced in analyzing olfactory and visual information. As was stated above, object recognition in visual space appears to be much more dependent on the analysis of local features (Lettvin et al., 1959; Marr, 1982) while olfactory object recognition appears to have no comparable sense of stimulus locality.

8. Possible Role of the Olfactory System in the Evolution of Cerebral Cortical Networks

To this point, the comparison of the visual and olfactory systems has emphasized differences in their structure. However, in many ways, one of the most striking aspects of the organization of these two systems is that, despite differences in the computation required of each, both employ the same basic circuit components. Thus, the olfactory and visual cortices, along with all other cerebral cortical networks, are fundamentally built around the properties of the pyramidal cell and its associated interneurons. In biology, this situation naturally leads to questions about the evolutionary significance of these shared features. This final section will briefly consider the possibility that the special relationship that appears to exist between olfactory pattern recognition and associative memory function could provide a key to understanding the development and evolution of cerebral cortex.

From a computational point of view, it seems at least plausible that the olfactory system could have served as the context for the initial evolution of the basic cerebral cortical associative memory network. As stated many times previously, the capacity to generate an associative network classifier seems fundamental to the most basic object recognition task performed by the olfactory system. This idea would also seem to be consistent with the particularly direct anatomical relationship between the olfactory periphery and the piriform cortex, as well as with the close relationship between piriform cortex and deep structures like the hippocampus. Viewed this way, the further evolution of the cerebral cortex may have resulted from the exploitation of this basic associative memory circuit by other sensory and motor systems. In keeping with the modeling work described in earlier sections of this chapter, the presumed importance of contextual information in olfactory object recognition may have served to promote the initial exposure of other sensory systems to the associative network. Further exploitation then presumably involved developing more elaborate preprocessing stages for sensory data arising from these other systems. Thus, the large number of distinct visual cortical regions, for example, can be seen as responsible for fashioning more complex representations of visual information for application to the associative memory. In this view then, the fundamental computational property of cerebral cortex is the ability to form context dependent associations that can serve to organize the interpretation and response to sensory information.

Unfortunately, it is not possible in this chapter to further consider these speculations. In fact, little relevant unambiguous data exits on these questions. However, it is worth pointing out that if these notions are correct, study of the olfactory system would seem a natural way to explore more general principles of cerebral cortical function (c.f. Haberly, 1985). Recently we have begun this process by comparing the behavior of our model of olfactory cortex with the same basic model modified to include features of primary visual cortex. Preliminary results suggest that these two different cortical structures may in fact share general computational properties (Wilson and Bower, 1990).

9. Conclusion and Caveats

This paper presents a very general overview of ongoing work in our laboratory designed to understand the relationship between the detailed structural organization of olfactory networks and the functions they perform. A series of computer simulations of the olfactory cortex of mammals have been briefly described that were designed to explore the possible mechanisms underlying object recognition in the olfactory system. The results show that a network based on the general structure of the olfactory cortex is capable of performing several associative functions that are likely to be important for olfactory processing.

While considerable progress has been made, it is important to realize that we are only at the very beginning of this effort. While the models described here are sophisticated by today's standards, they are at best rudimentary when judged against the true complexity of real cerebral cortical circuits. Undoubtedly, a great deal of additional information on the structural organization of these networks will need to be obtained before any detailed understanding of how they work can be reached. It is our conviction, however, that models of the type described here will play an essential role in this effort.

References

Bhalla, U.S., Wilson, M.A. & Bower, J.M. (1988). Integration of computer simulations and multi-unit recording in the rat olfactory system. *Soc. Neurosci. Abstr.* 14,1188.

Bower, J.M. (1990) Reverse engineering the nervous system: An anatomical, physiological, and computer based approach. In: S. Zornetzer, J. Davis, and C. Lau (Eds.) *An Introduction to Neural and Electronic Networks.* Academic Press pp. 3-24.

Bower, J.M. and Haberly, L.B. (1986). Facilitating and nonfacilitating synapses on pyramidal cells: A correlation between physiology and morphology. *Proc. Natl. Acad. Sci. USA.* 83: 1115-1119.

Bower, J.M., Nelson, M.E., Wilson, M.A., Fox, G.C. and Furmanski, W. (1988). Piriform (Olfactory) cortex model on the hypercube. In: G. Fox (Ed.), *Proceedings of 3rd conference on hypercube concurrent computers & applications.* ACM, New York, NY.

Bruce, C.J., Desimone, R. and Gross, C.G. (1977) Large receptive fields in a polysensory area in the superior temporal sulcus of the macaque. *Soc. Neurosci. Abstr.* 3:1756.

Cain, W.S. (1976) Physical and cognitive limitations on olfactory processing in human beings. In: D. Muller-Schwarze and M.M. Mozell (Eds.) *Chemical signals in vertebrates.* Plenum Press, New York. pp. 287-302.

Christensen, T.A. and Hildebrand, J.G. (1987). Male specific, sex pheromone-selective projection neurons in the antennal lobes of the moth Manduca Sexta. *J. Comp. Physiol.* 160, 553-569.

Desimone, R. and Gross, C.G. (1976) Absence of retinotopic organization in inferotemporal cortex. *Soc. Neurosci. Abstr.* 2:1108.

Desimone, R. and Gross, C.G. (1979) Visual areas in the temporal cortex of the macaque. *Brain Res.* 178:363-380.

Desimone, R. Albright, T.D., Gross, C.G. and Bruce, C. (1984) Stimulus-selective responses of inferior temporal neurons in the macaque. *J. Neurosci.* 4:2051-2062.

Devor, M. (1976). Fiber trajectories of olfactory bulb efferents in the hamster. *J. Comp. Neurol.* 166, 31-48.

Duchamp-Viret, P., Duchamp, A. and Vigouroux, M. (1989) Amplifying role of convergence in olfactory system. A comparative study of receptor cell and second-order neuron sensitivities. *J. Neurophysiol.* 61: 1085-1094.

Epple, G., Belcher, A., Greenfield, K.L., Kuderling, I., Nordstrom, K., & Smith, A.B. III (1989) Scent mixtures as social signals in two primate species: Saguinus fuscicollis and Saguinus o. oedipus. In: David G. Laing, William S, Cain, Robert L. McBride and Barry W. Ache (Eds.), *Perception of Complex Smells and Tastes.* New York, Academic Press, pp 1-26.

Freeman, W.J. (1975). *Mass Action in the Nervous System.* Academic Press, New York.

Gesteland R.C., Lettvin J.Y. and Pitts W.H. (1965) Chemical transmission in the nose of the frog. *J. Physiol (Lond.)* 181:525-559.

Gilbert, C.D. (1983) Microcircuitry of the visual cortex. *Ann Rev. Neurosci.* 6, 217-247.

Gross, C.G., Bender, D.G. and Rocha-Miranda, C.E. (1969) Visual receptive fields of neurons in inferotemporal cortex of the monkey. *Science* 166:1303-1306.

Gross, C.G., Rocha-Miranda, C.E. and Bender, D.B. (1972) Visual properties of neurons in inferotemporal cortex of the Macaque. *J. Neurophysiol.* 35:96-111.

Grossberg, S. (1988). *Neural Networks and Natural Intelligence.* MIT Press, Cambridge, Massachusetts.

Haberly, L.B. (1985). Neuronal circuitry in olfactory cortex: Anatomy and functional implications. *Chemical Senses* 10, 219-238.

Haberly, L.B. and Price, J.L. (1978) Association and commissural fiber systems of the olfactory cortex of the rat. I. Systems originating in the piriform cortex and adjacent areas. *J. Comp. Neurol.* 178:711-740.

Haberly, L.B. and J.M. Bower. (1989) Olfactory cortex: Model circuit for study of associative memory? *Trends in Neurosci.* 12: 258 - 264.

Hasselmo, M.E., Rolls, E.T., Baylis, G.C. and Nalwa, V. (1989a) Object-centered encoding by face-selective neurons in the cortex in the superior temporal sulcus of the monkey. *Exp. Brain Res.* 75:417-429.

Hasselmo, M.E., Rolls, E.T. and Baylis, G.C. (1989b) The role of expression and identity in the face-selective responses of neurons in the temporal visual cortex of the monkey. *Behav. Brain Res.* 32:203-218.

Hasselmo, M.E. and Bower, J.M. (1990) Afferent and association fiber differences in short-term potentiation in piriform (olfactory) cortex. *J. Neurophysiol.* (in press).

Hebb, D.O. (1949). *The Organization of Behavior*, Wiley, New York.

Hopfield, J.J. (1982). Neural networks and physical systems with emergent collective computational abilities. *Proc. Natl. Acad. Sci. USA.* 79, 2554-2558.

Hubel, D.H. and Wiesel, T.N. (1968) Receptive fields and functional architecture of monkey striate cortex. *J. Physiol.* 195:215-243.

Hubel, D.H. and Wiesel, T.N. (1972) Laminar and columnar distribution of geniculo-cortical fibers in the macaque monkey. *J. Comp. Neurol.* 146:421-450.

Hubel, D.H. (1982) Exploration of the primary visual cortex. *Nature* 299:516-525.

Insausti, R., Amaral, D.G. and Cowan, W.M. (1987) The entorhinal cortex of the monkey: II. Cortical afferents. *J. Comp. Neurol.* 264: 356-395.

Jastreboff, P.J., Pedersen, P.E., Greer, C.A., Stewart, W.B., Kauer, J.S., Benson, T.E. and Shepherd, G.M. (1984) Specific olfactory receptor populations projecting to identified glomeruli in the rat olfactory bulb. *Proc. Natl. Acad. Sci. USA* 81: 5250-5254.

Jones, E.G. and Powell, T.P.S. (1970) An anatomical study of converging sensory pathways within the cerebral cortex of the monkey. *Brain* 93:793-820

Kanter, E.D. and Haberly, L.B. (1989) APV dependent induction of long term potentiation in piriform (olfactory) cortex slices. *Soc. Neurosci. Abst.* 15: 929.

Kauer, J.S. (1987) Coding in the olfactory system. In: Finger T.E. and Silver W.L. (Eds.) *The Neurobiology of Taste and Smell.* Wiley, New York. pp. 205-231.

Kohonen, T. and Oja, E. (1976) Fast adaptive transformation of orthogonalizing filters and associative memory in recurrent networks of neuron-like elements. *Biol. Cyb.* 21, 85

Kohonen, T. (1984) *Self-organization and associative memory.* Springer-Verlag: Berlin.

Laing, D.G., William S, Cain, W.S., Robert L. McBride, R.L. and Barry W. Ache, B.W. (1989). *Perceptipn of Complex Smells and Tastes.* New York, Academic Press.

Lancet, D. (1986) Vertebrate olfactory reception. *Ann. Rev. Neurosci.* 9: 329-355.

Lettvin, J.Y. and Gestland, R.C. (1965). Speculations on smell. *Cold Spring Harbor Symp. Quant. Biol.* 30, 217-235.

Luskin, M.B. and Price, J.L. (1983) The topographic organization of associational fibers of the olfactory system in the rat, including centrifugal fibers to the olfactory bulb. *J. Comp. Neurol.* 216:264-291.

Marr, D. (1982) *Vision.* W.H. Freeman and Company, San Francisco.

Marr, D. and Hildreth, E. (1980) Theory of edge detection. *Proc. R. Soc. Lond. B* 207:187-217.

McNaughton, B.L. and Morris, R.G. (1987). Hippocampal synaptic enhancement and information storage within a distributed memory system. *Trends Neurosci.* 10, 408-415.

Mountcastle, V.B. (1957) Modality and topographic properties of single neurons of cat's somatic sensory cortex. *J. Neurophysiol* . 20:408-434.

Nelson. M.E. and Bower, J.M. (1990) Brain maps and parallel computers. *Trends in Neuroscience* (in press).

Nelson, M., Furmanski, W. and Bower, J.M. (1989). Simulating neural networks on parallel computers. In: C. Koch and I. Segev (Eds.) *Methods in Neuronal Modeling: From Synapses to Networks.* MIT Press, Cambridge, Massachusetts pp. 397-438.

Palm G. (1980) On associative memory. *Biol. Cybernetics* 36:19-31.

Perrett, D.I., Rolls, E.T. and Caan, W. (1982) Visual neurons responsive to faces in the monkey temporal cortex. *Exp. Brain Res.* 47:329-342.

Price, J.L. (1985) Beyond the primary olfactory cortex: olfactory related areas in the neocortex, thalamus and hypothalamus. *Chem. Senses* 10: 239-258.

Rabin, M.D., and Cain, W.S. (1989) Attention and learning in the perception of odor mixtures. In: D.G. Laing, S. William, W.S. Cain, L. Robert, R.L. McBride,and B. W. Ache (Eds) *Perceptipn of Complex Smells and Tastes.* New York, Academic Press. pp. 173-187.

Richardson, J.T.E. and Zucco, G.M. (1989) Cognition and olfaction: a review. *1989 Psychological bulletin* 105: 352-360.

Scott, J.W., McBride, R.L. and Schneider, S.P. (1980) The organization of projections from the olfactory bulb to the piriform cortex and olfactory tubercle in the rat. *J. Comp. Neurol.* 194: 519-534.

Sicard, G. and Holley, A. (1984) Receptor cell responses to odorants: Similarities and differences among odorants. *Brain Res.* 292: 283-296.

Squire, L.R. (1986a) Memory: Brain systems and behavior. *Trends Neurosci.* 11:170-175.

Squire, L.R. (1986b) Mechanisms of memory. *Science* 232:1612-1619.

Tanabe, T., Lino, M., and Takagi, S.F. (1975). Discrimination of odors in olfactory bulb, pyriform-amygdaloid areas and orbitofrontal cortex of the monkey. *J. Neurophysiol.* 38, 1284-1296.

Wigstrom, H., Gustafsson, B., Huang, Y.-Y. and Abraham, W.C. (1986) Hippocampal long-term potentiation is induced by pairing single afferent volleys with intracellularly injected depolarizing current pulses. *Acta Physiol. Scand.* 126:317-319.

Wilson, M. and Bower, J.M. (1988). A computer simulation of olfactory cortex with functional implications for storage and retrieval of olfactory information. In: D. Anderson (Ed.) *Neural information processing systems.* American Institute of Physics, New York. pp. 114-126.

Wilson, M. and Bower, J.M. (1989). The simulation of large-scale neuronal networks. In: C. Koch and I. Segev (Eds.) *Methods in Neuronal Modeling: From Synapses to Networks.* MIT Press, Cambridge, Massachusetts pp. 291-334.

5

CONTROL OF DYNAMIC SYSTEMS
VIA NEURAL NETWORKS

Stanislaw H. Zak, S. Mehdi Madani-Esfahani,
and Stefen Hui

1. INTRODUCTION

A neural network is a large-scale nonlinear circuit of interconnected simple circuits called nodes or neurons. These networks resemble patterns of biological neural networks hence the term neural networks. In fact motivation for studying such circuits came from attempts "to understand how known biophysical properties and (the) architectural organization of neural systems can provide the immense computational power characteristic of the brains of higher animals" (Tank and Hopfield, [24], p. 533). Another reason of a recent resurgence of interest in neural networks is the low execution speed of conventional computers which perform a program of instructions serially or sequentially. In contrast, neural networks operate in parallel. The ability to be interconnected in a regular fashion results in higher computation rates. Furthermore, regular interconnections of the same basic cells leads to easier design and testing of a chip.

Potential applications of neural networks are in such areas as speech and image recognition ([22], [28]), linear and nonlinear optimization ([3], [16], [24]), automatic control ([1], [2], [15]), and in highly parallel computers ([18]) to mention but a few.

The subject of this chapter is an application of additive neural network models to the control of dynamic processes. We begin with a brief description of a neural network model used in this chapter. Then we formulate the tracking problem and show how the additive neural network model can be used as a controller. A variable structure systems approach ([7], [26]) is utilized to construct the proposed controllers. We treat a class of neural networks considered in this chapter as variable structure systems with infinite gain [17]. This approach allows us to circumvent analysis problems caused by the discontinuous nonlinearity which is used to describe neurons. Finally, in the concluding Section we indicate directions for future research in the applications of neural networks to the control problems.

2. A BRIEF DESCRIPTION OF THE HOPFIELD NET

There are many neural network models ([11], [18]). In this chapter we will be concerned with the simpler additive model, also known as the Hopfield model, in the context of a control system tracking a reference signal. "The additive model has continued to be a cornerstone of neural network research to the present day... Some physicists unfamiliar with the classical status of the additive model in neural network theory erroneously called it the Hopfield model after they became acquainted with Hopfield's first application of the additive model in Hopfield (1984)" [14] - see Grossberg ([11], p. 23). The Hopfield neural network model is represented in Fig. 1.

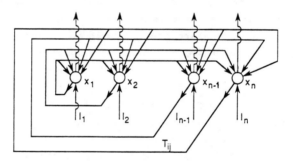

Fig. 1. Hopfield type neural network model.

Nodes or neurons, represented by circles in Fig. 1, can be modeled as shown in Fig. 2.

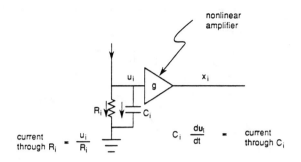

Fig. 2. Model of a neuron in the Hopfield net.

The nonlinear amplifier's input-output characteristic is described by the sigmoid function. The sigmoid function $x = g(u)$ ($g: \mathbb{R} \to \mathbb{R}$, where $\mathbb{R}$ denotes the set of real numbers) is defined by the properties:

(a) $|g(u)| \leq M$, where $0 < M < \infty$ is a constant, and

(b) $\dfrac{dg(u)}{du} \geq 0.$

A possible circuit implementation of the Hopfield network proposed by Smith and Portmann [23] is presented in Fig. 3. An equivalent representation of the circuit from Fig. 3. is given in Fig. 4.

This equivalent representation allows us to write down the equations governing the dynamical behavior of the net in a straightforward manner. Indeed, applying the Kirchoff current law at the input node of the amplifier and utilizing the fact that the input current into the amplifier is negligible (high input impedance) we obtain

$$C_i \frac{du_i}{dt} + \frac{u_i}{R_i} = \frac{x_1 - u_i}{R_{i1}} + \frac{x_2 - u_i}{R_{i2}} + ... + \frac{x_n - u_i}{R_{in}} + I_i, \quad i = 1, 2, ..., n \ . \quad (2.1)$$

Let

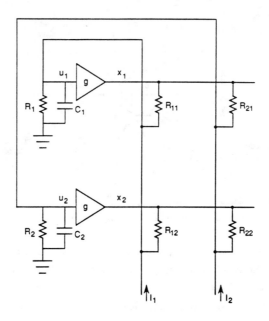

Fig. 3. Circuit realization of the Hopfield net.

$$T_{ij} = \frac{1}{R_{ij}} \; ,$$

$$\frac{1}{r_i} = \sum_{j=1}^{n} T_{ij} + \frac{1}{R_i} \; .$$

Utilizing the above notation we can represent equations (2.1) in the following form

$$C_i \frac{du_i}{dt} + \frac{u_i}{r_i} = \sum_{j=1}^{n} T_{ij} \, x_j + I_i \; , \quad i = 1,2,\dots n \; , \tag{2.2}$$

where $x_i = g(u_i)$. We can rewrite the equations (2.2) as

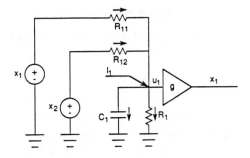

Fig. 4. Smith and Portmann's [23] equivalent representation of the
Hopfield network circuit realization.

$$\frac{du_i}{dt} = -b_i u_i + \sum_{j=1}^{n} A_{ij} \, g(u_j) + U_i \quad i = 1,2,...,n \qquad (2.3)$$

where

$$A_{ij} = \frac{T_{ij}}{C_i}, \quad U_i = U_i(t) = \frac{I_i(t)}{C_i}, \quad \text{and} \quad b_i = \frac{1}{r_i C_i}.$$

The qualitative analysis of the neural network model represented by (2.3) was performed among others, by Hopfield [14] and Michel et al. [21], see also [10] and [11].

The primary goal of this chapter is to show how the Hopfield type of neural networks can be applied to the control of dynamic processes, in particular to the tracking a reference signal by a control system.

The formulation of the tracking problem is the subject of the next Section.

3. FORMULATION OF THE TRACKING PROBLEM

Suppose we have a model of a dynamic process, plant, given by the following equations

$$\dot{x} = Fx + Gu \\ y = Hx \Biggr\}$$

(3.1)

where $F \in \mathbb{R}^{n \times n}$, $G \in \mathbb{R}^{n \times m}$, $u \in \mathbb{R}^m$, $y \in \mathbb{R}^m$, and $H \in \mathbb{R}^{m \times n}$. We wish to design a controller so that the closed-loop system can track a reference input with zero steady-state error (see Fig. 5). Let the reference signals be described, as in Davison [5], by the

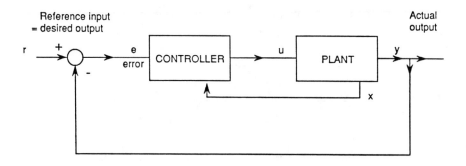

Fig. 5. Tracking system structure.

following differential equations

$$r_i^{(p)}(t) + \alpha_p r_i^{(p-1)}(t) + \dots + \alpha_2 \dot{r}_i(t) + \alpha_1 r_i(t) = 0 , \quad i = 1,2,\dots,m$$

(3.2)

where the initial conditions

$$r_i(0) , \quad \dot{r}_i(0) , \quad \dots , \quad r_i^{(p-1)}(0)$$

are specified.

Equations (3.2) can be rewritten in the form of the vector differential equation

$$r^{(p)}(t) + \alpha_p r^{(p-1)}(t) + \dots + \alpha_2 \dot{r}(t) + \alpha_1 r(t) = 0 ,$$

(3.3)

where $r(t) = [r_1(t),\dots,r_m(t)]^T \in \mathbb{R}^m$, $r^{(i)}(t) = [r_1^{(i)}(t),\dots,r_m^{(i)}(t)]^T \in \mathbb{R}^m$.

The tracking error is defined as

$$e(t) = y(t) - r(t) . \tag{3.4}$$

The problem of tracking $r(t) = [r_1(t), ..., r_m(t)]^T$ can be viewed as a designing exercise of a control strategy which provides regulation of the error (Franklin et al. [9], p. 390), that is, the error $e(t)$ should tend to zero as time gets large.

One way to solve this problem is to include the equations which are satisfied by the reference signal as a part of the control, (stabilization), problem in an error state space.

In particular, we differentiate the error equation p times and then introduce the error as a state. In what follows we extend results of Franklin et al. [9].

We have

$$
\begin{aligned}
e^{(p)} &= y^{(p)} - r^{(p)} \\
&= [\mathbf{H}x^{(p)} + \alpha_p r^{(p-1)} + \alpha_{p-1} r^{(p-2)} + ... + \alpha_1 r] \in \mathbb{R}^m .
\end{aligned} \tag{3.5}
$$

We then replace the plant state vector by the following vector

$$\xi = [x^{(p)} + \alpha_p x^{(p-1)} + ... + \alpha_1 x] \in \mathbb{R}^n \tag{3.6}$$

and define the new control vector as

$$v = [u^{(p)} + \alpha_p u^{(p-1)} + ... + \alpha_1 u] \in \mathbb{R}^m . \tag{3.7}$$

We now rewrite (3.5) as

$$
\begin{aligned}
e^{(p)} &= \alpha_p r^{(p-1)} - \alpha_p y^{(p-1)} + \alpha_p y^{(p-1)} \\
&\quad + ... + \alpha_1 r - \alpha_1 y + \alpha_1 y + \mathbf{H}x^{(p)} \\
&= - \alpha_p\, e^{(p-1)} - ... - \alpha_1 e + \mathbf{H}\left[x^{(p)} + \alpha_p x^{(p-1)} + ... + \alpha_1 x\right].
\end{aligned} \tag{3.8}
$$

Taking into account (3.6) yields

$$e^{(p)} = - \alpha_p e^{(p-1)} - ... - \alpha_1 e + \mathbf{H}\xi . \tag{3.9}$$

Differentiating ξ gives

$$\dot{\xi} = \mathbf{x}^{(p+1)} + \alpha_p \mathbf{x}^{(p)} + \dots + \alpha_1 \dot{\mathbf{x}}$$
$$= \mathbf{F}\mathbf{x}^{(p)} + \mathbf{G}\mathbf{u}^{(p)} + \alpha_p \mathbf{F}\mathbf{x}^{(p-1)} + \alpha_p \mathbf{G}\mathbf{u}^{(p-1)}$$
$$+ \dots + \alpha_1 \mathbf{F}\mathbf{x} + \alpha_1 \mathbf{G}\mathbf{u}$$
$$= \mathbf{F}(\mathbf{x}^{(p)} + \alpha_p \mathbf{x}^{(p-1)} + \dots + \alpha_1 \mathbf{x})$$
$$+ \mathbf{G}(\mathbf{u}^{(p)} + \alpha_p \mathbf{u}^{(p-1)} + \dots + \alpha_1 \mathbf{u}) . \tag{3.10}$$

Hence, in view of (3.7),

$$\dot{\xi} = \mathbf{F}\xi + \mathbf{G}\mathbf{v} . \tag{3.11}$$

Combining (3.9) and (3.11) yields

$$\begin{bmatrix} \dot{e} \\ \ddot{e} \\ \vdots \\ e^{(p-1)} \\ e^{(p)} \\ \dot{\xi} \end{bmatrix} = \begin{bmatrix} 0 & I_m & 0 & 0 & \dots & 0 & 0 \\ 0 & 0 & I_m & 0 & \dots & 0 & 0 \\ \vdots & \vdots & \vdots & \vdots & & \vdots & \vdots \\ 0 & 0 & 0 & 0 & \dots & I_m & 0 \\ -\alpha_1 I_m & \cdot & \cdot & \cdot & & -\alpha_p I_m & H \\ 0 & 0 & 0 & 0 & & 0 & F \end{bmatrix} \begin{bmatrix} e \\ \dot{e} \\ \vdots \\ e^{(p-2)} \\ e^{(p-1)} \\ \xi \end{bmatrix} + \begin{bmatrix} 0 \\ 0 \\ \vdots \\ 0 \\ 0 \\ G \end{bmatrix} \mathbf{v} . \tag{3.12}$$

Let $z^T \triangleq [e^T, \dot{e}^T, \dots, \xi^T] \in \mathbb{R}^{1 \times (pm+n)}$,

$$A = \begin{bmatrix} 0 & I_m & \dots & 0 & 0 \\ \vdots & & & & \vdots \\ -\alpha_1 I_m & & \dots & -\alpha_p I_m & H \\ 0 & & \dots & 0 & F \end{bmatrix}, \quad B = \begin{bmatrix} 0 \\ \vdots \\ 0 \\ G \end{bmatrix} .$$

Then (3.12) can be represented as

$$\dot{z} = Az + Bv . \tag{3.13}$$

We refer to the system modeled by (3.13) as the error system. *Thus the problem of the regulation of the error is reduced to the problem of the stabilization of the error system.* Different approaches to the stabilization problem of dynamic systems modeled by (3.13) exist. For example see DeCarlo [6] or Franklin et al. [9]. We propose two new approaches to designing stabilizing control strategies for (3.13) involving the Hopfield type neural networks. Prior to presenting these control laws we will introduce the necessary apparatus for further analysis. This is the subject of the following Section.

4. AUXILIARY RESULTS

In the following analysis we utilize certain concepts from the theory of variable structure control. Let us now look at the equations (2.3) which describe the Hopfield neural network model. Assume that

$$g(u_i) = -\text{sgn}(u_i) = \begin{cases} -1 & \text{if} & u_i > 0 \\ 0 & \text{if} & u_i = 0 \\ +1 & \text{if} & u_i < 0 . \end{cases}$$

The above assumption implies that the nonlinear amplifiers have infinite gains. Under such an assumption, the right hand side of (2.3) is discontinuous. The system (2.3) may be referred to as a variable structure system [17].

Variable Structure Control (VSC), the control of dynamical systems with discontinuous state feedback controllers, has been developed over the last 25 years. See [7], and [26] for surveys and [12], [25], [29] for applications. This theory rests on the concept of changing the structure of the controller in response to the changing states of the system to obtain a desired response. This is accomplished by the use of a high speed switching control law which forces the trajectories of the system onto a chosen manifold, where they are maintained thereafter. The system is insensitive to certain parameter variations and disturbances while the trajectories are on the manifold. If the state vector is not accessible, then a suitable estimate must be used.

We use the following notation. If $x \in \mathbb{R}^n$, then $\|x\|$ denotes the Euclidean norm, that is, $\|x\| = (x_1^2 + \dots + x_n^2)^{1/2}$. If A is a matrix, then $\|A\|$ is the spectral norm defined by

$$\|A\| = \max_x \{\|Ax\| \mid \|x\| \leq 1\} .$$

For any square matrix A, we let $\lambda_{\min}(A)$ be the minimum eigenvalue of A and $\lambda_{\max}(A)$ be the maximum eigenvalue of A.

With this notation,

$$\|A\|^2 = \lambda_{\max}(A^T A) .$$

Rayleigh principle states that if P is a real symmetric positive definite matrix, then

$$\lambda_{\min}(P) \|x\|^2 \leq x^T P x \leq \lambda_{\max}(P) \|x\|^2 .$$

An important concept in variable structure control is that of an attractive manifold on which certain desired dynamical behavior is guaranteed. Trajectories of the system should be steered towards the manifold and subsequently constrained to remain on it.

We next describe the manifold which is used in this chapter. Suppose

$$S = \begin{bmatrix} s_1 \\ \vdots \\ s_m \end{bmatrix} \in \mathbb{R}^{m \times n} ,$$

where

$$s_i \in \mathbb{R}^{1 \times n} .$$

We assume that S is of full rank.

Let

$$\sigma(x) = \begin{bmatrix} \sigma_1(x) \\ \vdots \\ \sigma_m(x) \end{bmatrix} = \begin{bmatrix} s_1 x \\ \vdots \\ s_m x \end{bmatrix} = Sx ,$$

where

$$x \in \mathbb{R}^n ,$$

and let

$$\Omega = \{x \mid \sigma(x) = 0\} .$$

Consider the system

$$\dot{x} = Ax + Bu , \tag{4.1}$$

where $x \in \mathbb{R}^n$, $A \in \mathbb{R}^{n \times n}$, $B \in \mathbb{R}^{n \times m}$, $u \in \mathbb{R}^m$. We make the following assumptions:

Assumption 1: The matrix SB is nonsingular.

Assumption 2: The pair (A,B) is completely controllable.

Definition 4.1. ([26]) The solution of the algebraic equation in u of

$$S\dot{x} = SAx + SBu = 0$$

is called the equivalent control and denoted by u_{eq}, that is,

$$u_{eq} = -(SB)^{-1}SAx .$$

Definition 4.2. The equivalent system is the system that is obtained when the original control u is replaced by the equivalent control u_{eq}, that is,

$$\dot{x} = [I_n - B(SB)^{-1}S]A \, x.$$

We assume that the control u in the system (4.1) is bounded and that

$$|u_i| \leq \mu_i , \quad i = 1,...,m , \tag{4.2}$$

where $\mu_i > 0$.

Our goal is to design a controller which satisfies the bounds (4.2) and which induces the sliding mode on Ω in the sense of Definition 4.3.

Definition 4.3. ([26])

A domain Δ in the manifold $\{x \mid \sigma(x) = 0\}$ is a sliding mode domain if for each $\epsilon > 0$ there exists a $\delta > 0$ such that any trajectory starting in the n-dimensional δ-neighborhood of Δ may leave the n-dimensional ϵ-neighborhood of Δ only through the n-dimensional ϵ-neighborhood of the boundary of Δ (see Fig. 6).

In general, the controller in VSC varies its structure depending on the position relative to the switching surface and has the form [26]:

$$u_i = \begin{cases} u_i^+(x) & \text{if} \quad \sigma_i(x) > 0 \\ u_i^-(x) & \text{if} \quad \sigma_i(x) < 0 . \end{cases}$$

It is easy to see that if $\sigma^T \dot{\sigma} < 0$, then the trajectory is tending towards the switching surface. Hence if $\sigma^T \dot{\sigma} < 0$ in a neighborhood of a region Δ of the switching surface, then Δ is a sliding domain ([26]). For example, let

$$\tilde{u} = \begin{bmatrix} -k_1 \operatorname{sgn} \sigma_1 \\ \vdots \\ -k_m \operatorname{sgn} \sigma_m \end{bmatrix}$$

where

111

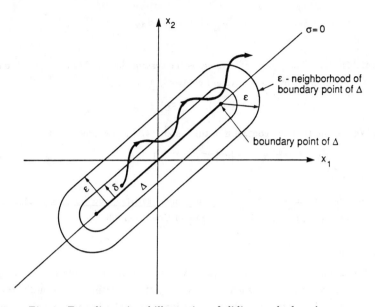

Fig. 6. Two-dimensional illustration of sliding mode domain.

$$\text{sgn } \sigma_i = \begin{cases} 1 & \text{if} \quad \sigma_i > 0 \\ 0 & \text{if} \quad \sigma_i = 0 \\ -1 & \text{if} \quad \sigma_i < 0 \,, \end{cases}$$

and $k_i > 0$ for $i = 1,...,m$. One can easily check that if

$$u = u_{eq} + (SB)^{-1}\tilde{u} \,,$$

then

$$\sigma^T \dot{\sigma} = \sigma^T \tilde{u} < 0 \,.$$

Hence with the above control u, we have a sliding mode.

Let us assume that the switching surface is chosen so that $SB = I_m$; we will see how this can be accomplished in Sections 6 and 8. With this assumption, $u_{eq} = -SAx$. We next give a sufficient condition for $\sigma^T \dot{\sigma} < 0$ to hold. Note that

$$\sigma^T \dot{\sigma} = \sigma^T(SAx + SBu)$$
$$= \sigma^T(-u_{eq} + u)$$
$$= \sum_{i=1}^{m} \sigma_i(-(u_{eq})_i + u_i) .$$

Hence if

$$\begin{cases} u_i^+(x) < (u_{eq})_i & \text{for} \quad \sigma_i(x) > 0 \\ u_i^-(x) > (u_{eq})_i & \text{for} \quad \sigma_i(x) < 0 , \end{cases} \tag{4.3}$$

then $\sigma^T \dot{\sigma} < 0$. In this chapter, we use the control law

$$u = \begin{bmatrix} -\mu_1 \text{sgn} \sigma_1 \\ \vdots \\ -\mu_m \text{sgn} \sigma_m \end{bmatrix} .$$

With this control law, we have $\sigma^T \dot{\sigma} < 0$ in the region

$$\Omega^* = \bigcap_{i=1}^{m} \{x \mid |(u_{eq})_i| < \mu_i\}$$
$$= \bigcap_{i=1}^{m} \{x \mid |s_i Ax| < \mu_i\} .$$

Note that Ω^* is an open neighborhood of the origin. If $\Delta = \Omega \cap \Omega^*$, where $\Omega = \{x \mid \sigma(x) = 0\}$, then Δ is a sliding domain.

Observe that a sliding domain is a region of asymptotic stability (RAS) for the system.

Example 4.1. Consider the dynamical system modeled by:

$$\ddot{x} = u .$$

In the state-space representation, we have

$$\begin{bmatrix} \dot{x}_1 \\ \dot{x}_2 \end{bmatrix} = \begin{bmatrix} 0 & 1 \\ 0 & 0 \end{bmatrix} \begin{bmatrix} x_1 \\ x_2 \end{bmatrix} + \begin{bmatrix} 0 \\ 1 \end{bmatrix} u .$$

Let $u = -\text{sgn } \sigma(x)$, and $\sigma(x) = x_1 + x_2 = [1 \ 1]x$. In this case

$$\Omega = \{(x_1, x_2) \mid x_1 + x_2 = 0\},$$
$$\Omega^* = \{(x_1, x_2) \mid |x_2| < 1\}$$

and $\Delta = \Omega^* \cap \Omega$ is the segment of the line $x_1 + x_2 = 0$ with $|x_2| < 1$ (see Fig. 7).

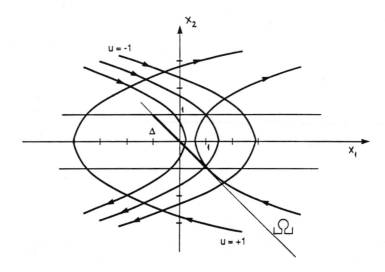

Fig. 7. Phase-plane portrait and illustration of a sliding domain in Example 4.1.

5. CONTROL OF SINGLE-INPUT SYSTEMS

There are two basic steps in the design of VSC:

(1) The design of the switching surface (manifold) so that the behavior of the system has certain prescribed properties on the surface. For example, the switching surface will be designed so that the system is asymptotically stable on the surface.

(2) The design of the control strategy to steer the system to the switching surface and to maintain it there.

In this Section we consider a class of single-input dynamic system modeled by

$$\dot{x} = Ax + bu \qquad (5.1)$$

where $x \in \mathbb{R}^n$, $A \in \mathbb{R}^{n \times n}$, $b \in \mathbb{R}^{n \times 1}$, $u \in \mathbb{R}$. We assume that the pair (A,b) is completely controllable and hence (5.1) is equivalent to the controller canonical form

114

$$\dot{x} = \begin{bmatrix} 0 & 1 & 0 & 0 & 0 & 0 \\ 0 & 0 & 1 & 0 & 0 & 0 \\ \vdots & \vdots & & & \vdots & \vdots \\ 0 & 0 & \dots & & 0 & 1 \\ \alpha_1 & \alpha_2 & \dots & & \alpha_{n-1} & \alpha_n \end{bmatrix} x + \begin{bmatrix} 0 \\ 0 \\ \vdots \\ 0 \\ 1 \end{bmatrix} u . \tag{5.2}$$

The mainfold we use has the form

$$\Omega = \{ x \mid sx = 0 \} , \quad s \in \mathbb{R}^{1 \times n} . \tag{5.3}$$

We can assume that $s = [s_1, \dots, s_{n-1}, 1]$, $s_i \in \mathbb{R}$. Observe that if the system is in the controller canonical form then $sb = 1$. When the system dynamics is given by the controller canonical form then the equivalent system is

$$\dot{x} = \begin{bmatrix} 0 & 1 & 0 & \dots & 0 & 0 \\ 0 & 0 & 1 & \dots & 0 & 0 \\ \vdots & & & & & \\ 0 & 0 & \dots & \dots & 0 & 1 \\ 0 & -s_1 & -s_2 & \dots & -s_{n-2} & -s_{n-1} \end{bmatrix} x . \tag{5.4}$$

The controller on which we will concentrate is

$$u = - \mu \mathrm{sgn}(sx) , \tag{5.5}$$

where μ is a positive real number. Note that this controller is bounded by μ.

We now choose the switching surface so that the system restricted to this surface has prescribed distinct negative eigenvalues $-\lambda_1, \dots, -\lambda_{n-1}$, $\lambda_i > 0$, $i = 1, \dots, n-1$. If the system is in the controller canonical form, then in sliding mode the system is described by (5.4) and $sx = 0$. The order of the system in sliding is n-1 and its characteristic equation is given by

$$\lambda^{n-1} + s_{n-1} \lambda^{n-2} + \dots + s_1 = 0 . \tag{5.6}$$

The prescribed eigenvalues $-\lambda_1, \dots, -\lambda_{n-1}$ must satisfy (5.6) and hence we have the linear equations

$$\begin{bmatrix} 1 & -\lambda_1 & (-\lambda_1)^2 & \dots & (-\lambda_1)^{n-1} \\ & \vdots & & & \\ 1 & -\lambda_{n-1} & (-\lambda_{n-1})^2 & \dots & (-\lambda_{n-1})^{n-1} \end{bmatrix} \begin{bmatrix} s_1 \\ \vdots \\ s_{n-1} \\ 1 \end{bmatrix} = 0 . \tag{5.7}$$

Since $\lambda_1, \dots, \lambda_{n-1}$ are distinct, the coefficient matrix has full rank and $s_1, \dots, s_{n-1}$ are uniquely determined. This completes the design of the switching surface

$\Omega = \{x \mid sx = 0\}$. Note that one does not have to have a model of a dynamic system to be controlled in the controller canonical form. We have used the canonical form to facilitate the analysis.

To proceed further we introduce a state-variable transformation. For ℓ, k positive integers, we let

$$V_\ell(\beta_1,...,\beta_k) = \begin{bmatrix} 1 & 1 & \cdots & 1 \\ \beta_1 & \beta_2 & & \beta_k \\ \vdots & \vdots & & \vdots \\ \beta_1^\ell & \beta_2^\ell & & \beta_k^\ell \end{bmatrix} \in \mathbb{R}^{(\ell+1)\times k} .$$

Let

$$W = V_{n-1}(-\lambda_1,...,-\lambda_{n-1}) \in \mathbb{R}^{n\times(n-1)} ,$$

and

$$W^g = [\{V_{n-2}(-\lambda_1,...,-\lambda_{n-1})\}^{-1} \mid 0] \in \mathbb{R}^{(n-1)\times n} .$$

Note that $W^g W = I_{n-1}$.

Let, as in [20],

$$M = \begin{bmatrix} W^g \\ s \end{bmatrix} .$$

Observe that $M^{-1} = [W \ b]$ (see (5.7)). We introduce the new coordinates

$$\begin{bmatrix} z \\ y \end{bmatrix} = Mx = \begin{bmatrix} W^g x \\ sx \end{bmatrix} , \tag{5.8}$$

where $z \in \mathbb{R}^{n-1}$, $y \in \mathbb{R}$. In these coordinates, the system (5.2) has the form

$$\begin{aligned} \dot{z} &= A_{11} z + A_{12} y \\ \dot{y} &= A_{21} z + A_{22} y + u , \end{aligned} \tag{5.9}$$

where $A_{11} = \text{diag}(-\lambda_1,...,-\lambda_{n-1})$. If we use the controller

$$u = -\mu \, \text{sgn}(sx) = -\mu \, \text{sgn} \, y ,$$

then the system (5.2) in the new coordinates is described by

$$\begin{aligned} \dot{z} &= A_{11} z + A_{12} y \\ \dot{y} &= A_{21} z + A_{22} y - \mu \, \text{sgn} \, y . \end{aligned} \tag{5.10}$$

Remark 5.1.

One can interpret equations (5.10) as follows. We are given a dynamic system

$$\dot{z} = A_{11}z + A_{12}y \qquad (5.11)$$

driven by or connected to a single neuron type controller

$$\dot{y} = A_{22}y - \mu\,\mathrm{sgn}y + A_{21}z. \qquad (5.12)$$

This observation follows from the comparison of (2.3) and (5.12), where $U = A_{21}z$.

Note that in order to arrive at the above conclusion we had to perform a state-space transformation (5.8) to reveal the implicit presence of the neural controller in the closed-loop system (5.2), (5.5).

Remark 5.2.

We can explicitly employ the neural type controller to a given dynamic system. In particular suppose we are given a dynamic system modeled by (5.1). We propose a controller of the form (2.3)

$$\dot{u} = -\beta u - \mu\,\mathrm{sgn}\,\sigma(x,u) + U\,, \qquad (5.13)$$

where $U = c^T x$, and $\sigma(x,u)$ is a switching surface to be chosen. The equations of the closed-loop system can be represented as

$$\begin{bmatrix} \dot{x} \\ \dot{u} \end{bmatrix} = \begin{bmatrix} A & b \\ c^T & -\beta \end{bmatrix} \begin{bmatrix} x \\ u \end{bmatrix} + \begin{bmatrix} 0 \\ 1 \end{bmatrix} (-\mu\,\mathrm{sgn}\,\sigma(x,u))\,. \qquad (5.14)$$

Observe that $[x^T, u]^T \in \mathbb{R}^{n+1}$. In order to proceed further one has to decide what kind of dynamic behavior is to be imposed on the closed-loop system. This then should be expressed in the form of n prescribed eigenvalues which will correspond to the eigenvalues of the system while in sliding along $\sigma(x,u) = 0$. Having chosen desired eigenvalues we can determine the switching surface $\sigma(x,u)$ using (5.7). If one then transforms (5.14) first into the controller canonical form and then into the new coordinates utilizing (5.8), then the resulting system will be in the form (5.10).

Example 5.1. Consider the system given by

$$\begin{bmatrix} \dot{x}_1 \\ \dot{x}_2 \end{bmatrix} = \begin{bmatrix} 0 & 1 \\ 0 & 0 \end{bmatrix} \begin{bmatrix} x_1 \\ x_2 \end{bmatrix} + \begin{bmatrix} 0 \\ 1 \end{bmatrix} u , \qquad (5.15)$$

where $u = -\mu \, \mathrm{sgn} \, \sigma(x)$. We choose the switching surface so that the system restricted to the switching surface has eigenvalue $-\lambda_1$. From (5.6) and (5.7) the switching surface is

$$s_1 x_1 + x_2 = \lambda_1 x_1 + x_2 = 0 .$$

We use the controller

$$u = -\mu \, \mathrm{sgn}(\lambda_1 x_1 + x_2) . \qquad (5.16)$$

A block diagram representation of the closed-loop system is given in Fig. 8. Using the

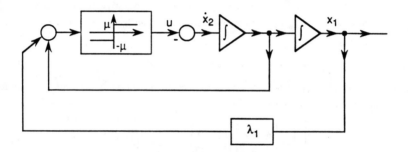

Fig. 8. Block diagram of the closed-loop system (5.15), (5.16)
in Example 5.1 in the old coordinates.

method described above we have

$$M = \begin{bmatrix} W^g \\ s \end{bmatrix} = \begin{bmatrix} 1 & 0 \\ \lambda_1 & 1 \end{bmatrix},$$

and the system in the new coordinates is

$$\dot{z} = -\lambda_1 z + y$$
$$\dot{y} = -\lambda_1^2 z + \lambda_1 y - \mu \, \mathrm{sgn} \, y .$$

Consider again the system (5.15). This time we choose an explict neural controller

$$\dot{u} = -\beta u + c^T x - \mu \operatorname{sgn} \sigma(x, u), \tag{5.17}$$

where $c^T = [c_1, c_2]$. The closed-loop system now has the form

$$\begin{bmatrix} \dot{x}_1 \\ \dot{x}_2 \\ \dot{u} \end{bmatrix} = \begin{bmatrix} 0 & 1 & 0 \\ 0 & 0 & 1 \\ c_1 & c_2 & -\beta \end{bmatrix} \begin{bmatrix} x_1 \\ x_2 \\ u \end{bmatrix} + \begin{bmatrix} 0 \\ 0 \\ 1 \end{bmatrix} (-\mu \operatorname{sgn} \sigma(x, u)) \tag{5.18}$$

A block diagram of this closed-loop system is depicted in Fig. 9. The switching surface

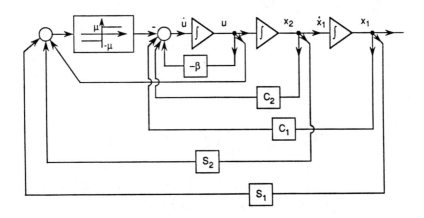

Fig. 9. Block diagram of the closed-loop system (5.18).

is found from equation (5.7)

$$\begin{bmatrix} 1 & -\lambda_1 & (-\lambda_1)^2 \\ 1 & -\lambda_2 & (-\lambda_2)^2 \end{bmatrix} \begin{bmatrix} s_1 \\ s_2 \\ 1 \end{bmatrix} = 0, \tag{5.19}$$

where $-\lambda_1$ and $-\lambda_2$ are the desired (distinct) eigenvalues. Solving (5.19) yields

$$\begin{bmatrix} s_1 \\ s_2 \end{bmatrix} = \frac{1}{\lambda_2 - \lambda_1} \begin{bmatrix} -\lambda_2 & \lambda_1 \\ -1 & 1 \end{bmatrix} \begin{bmatrix} \lambda_1^2 \\ \lambda_2^2 \end{bmatrix} = \begin{bmatrix} \lambda_1 \lambda_2 \\ \lambda_2 + \lambda_1 \end{bmatrix}. \tag{5.20}$$

We now can construct the state-space transformation (5.8). We have

$$M = \begin{bmatrix} V_1^{-1} & | & 0 \\ & | & 0 \\ -- & -- & -- \\ & s & \end{bmatrix} = \begin{bmatrix} W^g \\ s \end{bmatrix} = \begin{bmatrix} \dfrac{\lambda_2}{\lambda_2 - \lambda_1} & \dfrac{1}{\lambda_2 - \lambda_1} & 0 \\[2mm] \dfrac{\lambda_1}{\lambda_1 - \lambda_2} & \dfrac{1}{\lambda_1 - \lambda_2} & 0 \\[2mm] \lambda_1 \lambda_2 & \lambda_2 + \lambda_1 & 1 \end{bmatrix}.$$

Note that

$$M^{-1} = [W \ \ b] = [V_2 \ \vdots \ b]$$

$$= \begin{bmatrix} 1 & 1 & 0 \\ -\lambda_1 & -\lambda_2 & 0 \\ (-\lambda_1)^2 & (-\lambda_2)^2 & 1 \end{bmatrix}.$$

In the new coordinates (5.18) has the form (5.10) where

$$A_{11} = \begin{bmatrix} -\lambda_1 & 0 \\ 0 & -\lambda_2 \end{bmatrix} \quad \text{and} \quad y = [s_1 \ \ s_2 \ \ 1] \begin{bmatrix} x_1 \\ x_2 \\ u \end{bmatrix}.$$

The results of Section 4 imply that the closed-loop system described by (5.10) is locally asymptotically stable. The goal of the next Section is to investigate regions of asymptotic stability for dynamic systems driven by or connected to a single neuron type controller. In particular, we shall be interested in the region of asymptotic stability of the closed-loop system modeled by (5.14). We shall also investigate the way the systems trajectories reach the equilibrium point $x_{eq} = 0$. Specifically, we will give estimates of the sliding mode domain of a dynamic system driven by a neural type controller. To facilitate the analysis we will be working in appropriately chosen coordinates, namely we shall employ the necessary state-space transformation which brings the closed-loop system into the form described by (5.10).

6. A FIRST APPROXIMATION OF THE REGION OF ASYMPTOTIC STABILITY WITH SLIDING

In this Section we give a first approximation of the region of asymptotic stability (RAS) with sliding. The results of this Section are based on the paper by Madani-Esfahani et al. [20].

Better approximation will be given in the following Section. We start with the following lemma.

The system under consideration is modeled by (5.10).

Lemma 6.1. ([20]). For $0 < \epsilon < \mu$, the region

$$\Delta = \{(z,0) \mid |A_{21}z| < \mu - \epsilon\} \subset \Omega$$

is a sliding mode domain (see Definition 4.3).

Proof: Let $\Delta_\epsilon = \{(z,y) \mid |A_{21}z| < \mu - \epsilon, |y| < \dfrac{\epsilon}{|A_{22}|}\}$.   Then   in

$\Delta_\epsilon \setminus \{(z,y) \mid y = 0\}$

$$\frac{1}{2} \frac{d}{dt}(y^2) = y\,\dot{y} = y(A_{21}z + A_{22}y) - \mu\,|y|$$

$$< |y|(\mu - \epsilon + \epsilon - \mu) = 0 .$$

Therefore a trajectory starting in Δ_ϵ can leave Δ_ϵ only through the $\epsilon/|A_{22}|$-neighborhood of the boundary of Δ in Ω.

$\square$

Observe that if the initial point is in Δ, then the system will be in sliding for some positive time. However, there is no guarantee that we stay in Δ for subsequent times. From the fact that $A_{11} = \text{diag}\,(-\lambda_1, ..., -\lambda_{n-1})$ we have the following. Let B_r denote the ball centered at 0 with radius r.

Proposition 6.2. ([20]). Let

$$R = \sup\,\{r \mid B_r \cap \Omega \subset \Delta\},$$

and

121

$$\Sigma = B_R \cap \Omega.$$

Then Σ is a RAS with sliding.

Proof: While in sliding the system is governed by $\dot{z} = \text{diag} (-\lambda_1,...,-\lambda_{n-1})z$. Hence $z_i = z_i(0)e^{-\lambda_i t}$ for $i = 1,...,n-1$. Thus if $\sum_{i=1}^{n-1} z_i^2(0) < R^2$, then $\sum_{i=1}^{n-1} z_i^2(t) < R^2$ and by Lemma 6.1 $z(t) \in \Sigma$ for $t \geq 0$.

□

Note that Σ is the largest circular region that is contained in Δ. We can easily see that $R = \mu/a_{21}$, where $a_{21} = \|A_{21}\|$, the Euclidean norm of A_{21}.

7. IMPROVED ESTIMATES OF RAS WITH SLIDING

In the previous Section we obtained a RAS with sliding contained in the switching surface. We now use this information to obtain RAS's that are not constrained to the switching surface. The method we use is that of finding RAS's whose restriction to the switching surface is contained in Σ, and hence will be a RAS with sliding. Our main tool is a Lure-like Lyapunov function candidate

$$V(z,y;\beta,\eta,h) = (a_{21}\|z\|)^2 + 2\beta(A_{21}z)y + hy^2 + \mu\eta |y|, \tag{7.1}$$

where β, h, and η are positive constants. When there is no ambiguity, we will write $V(z,y)$ for $V(z,y;\beta,\eta,h)$. Observe that

$$V(z,y) \geq (a_{21}\|z\| - \beta |y|)^2 + (h - \beta^2)y^2 + \mu\eta |y|. \tag{7.2}$$

Hence if $h - \beta^2 \geq 0$, then V is positive in $\mathbb{R}^n \backslash 0$. If $\beta^2 - h > 0$ then V is positive in $\{(z,y) \mid |y| < \dfrac{\mu\eta}{\beta^2 - h}\} \backslash 0$. Since V contains a multiple of $|y|$, the Lyapunov derivative $\dot{V}$ may not exist on a trajectory which intersects $\Omega = \{(z,y) \mid y = 0\}$. However, when restricted to Ω the system takes the form $\dot{z} = A_{11}z = \text{diag} (-\lambda_1,...,-\lambda_{n-1})z$. Therefore if the trajectory $(z(t),y(t))$ is in Ω for $t_1 \leq t \leq t_2$, we must have $\|z(t_2)\| \leq \|z(t_1)\|$. Since the restriction of level sets of V to Ω are circular regions, the trajectories of the system cannot leave a sublevel set $\{(z,y) \mid V(z,y) < a^2, a \in \mathbb{R}\}$ of V through Ω. Therefore if Γ is a region such that

(i) V is positive in Γ,

(ii) $\dot{V}$ is negative in $\Gamma\backslash\Omega$,

then the largest sublevel set of V contained in Γ is a RAS. If in addition we have

(iii) the restriction of Γ to Ω is contained in Σ,

then the largest sublevel set of V contained in Γ is a RAS with sliding. Note that we do not need to consider $\dot{V}$ on Ω.

Theorem 7.1. Suppose $\eta \geq 2\beta$, $a_{21} \neq 0$. Then there exist positive constants $C_1 = C_1(A,\mu,\beta,\eta)$ and $C_2 = C_2(A,\mu,\beta,\eta)$ such that for $\delta \in (0,\mu)$ and

$$h \geq \max\left\{\frac{C_1}{\delta^2} + \beta^2 , \ \frac{C_2}{\delta}\right\} \text{the region}$$

$$\{(z,y)\,|V(z,y;\eta,\beta,h) < (\mu - \delta)^2\}$$

is a RAS with sliding.

Proof: Suppose $0 < \delta < \mu$. Let

$$\Gamma = \{(z,y)\,|V(z,y) < (\mu - \delta)^2\} \,.$$

From the discussion before the theorem we need to show that

(i) $V > 0$ in Γ,

(ii) $\dot{V} < 0$ in $\Gamma\backslash\Omega$,

(iii) $\Gamma\cap\Omega\subset\Sigma$.

Condition (i) holds since by assumption $h \geq \max\left\{\dfrac{C_1}{\delta_2} + \beta^2 , \ \dfrac{C_2}{\delta}\right\}$, and thus $h > \beta^2$ which implies that V is positive definite.

Condition (ii) holds by definition. Indeed $V(z,0) = (a_{21}\|z\|)^2 < (\mu - \delta)^2$ implies $a_{21}\|z\| < \mu - \delta$. Hence by Lemma 6.1 $\Gamma\cap\Omega\subset\Sigma$.

We need to prove (iii). After some manipulations one obtains

$$\dot{V} = 2a_{21}^2 z^T A_{11} z + \left[2\beta(A_{21}z)^2 - \eta\mu^2\right]$$

$$+ 2\,h\,|y|\left[A_{21}z\,\operatorname{sgn}\,y - \mu\right] + \mu A_{21}z\left[\eta - 2\beta\right]\operatorname{sgn}\,y$$

$$+ A_{22}\,|y|\left[2h\,|y| + 2\beta(A_{21}z)\,\operatorname{sgn}\,y + \mu\eta\right]$$

$$+ 2y\left[\beta A_{21}A_{11}z + \beta y A_{21}A_{12} + a_{21}^2 z^T A_{12}\right].$$

Choose $\tau > 0$ so that $\beta\tau < \dfrac{1}{4}$ and $a_{22}\tau < \dfrac{1}{4}$. Let $(z,y)\in\Gamma$. If $h \geq \left[\dfrac{\mu-\delta}{\tau\delta}\right]^2 + \beta^2$ we see from (7.2) that $|y| < \tau\delta$. Indeed $(z,y)\in\Gamma$ implies

$$(a_{21}||z|| - \beta\,|y|)^2 + (h - \beta^2)y^2 + \mu\eta\,|y| < (\mu - \delta)^2\,.$$

If $h \geq \left[\dfrac{\mu-\delta}{\tau\delta}\right]^2 + \beta^2$ then

$$\left[\dfrac{\mu-\delta}{\tau\delta}\right]^2 y^2 < (a_{21}||z|| - \beta\,|y|)^2 + \left[\dfrac{\mu-\delta}{\tau\delta}\right]^2 y^2 + \mu\eta\,|y| < (\mu - \delta)^2\,.$$

Therefore

$$y < \tau\delta\,.$$

If in addition $\beta\tau < \dfrac{1}{4}$ then

$$(a_{21}||z|| - \beta\,|y|)^2 < (a_{21}||z|| - \beta\,|y|)^2 + \left[\dfrac{\mu-\delta}{\tau\delta}\right]^2 y^2 + \mu\eta\,|y| < (\mu - \delta)^2$$

and hence

$$a_{21}||z|| - \beta\,|y| < \mu - \delta$$

implies

$$a_{21}||z|| < \mu - \dfrac{3}{4}\,\delta\,.$$

Since $\eta \geq 2\beta$ and $|A_{21}z| < \mu$ (see Lemma 6.1), we have

$$[2\beta(A_{21}z)^2 - \eta\mu^2] + \mu A_{21}z[\eta - 2\beta]\,\operatorname{sgn}\,y$$

$$\leq 2\beta(A_{21}z)^2 - \eta\mu^2 + \mu \,|A_{21}z| \,[\eta - 2\beta]$$

$$= (2\beta \,|A_{21}z| + \eta\mu)(\,|A_{21}z| - \mu) < 0 \,.$$

This combined with the fact that $A_{11} = \text{diag}(-\lambda_1, ..., -\lambda_{n-1})$, $-\lambda_i < 0$, and $\|z\| < \left(\mu - \dfrac{3}{4}\,\delta\right)/a_{21} \leq \mu/a_{21}$ shows that one can find a constant $K = K(A, \mu, \eta, \beta)$ such that

$$\dot{V} < 2h\,|y|\,(\,|A_{21}z| - \mu + a_{22}\,|y|) + K(\,|y| + |y|^2) \,.$$

Since $|y| < \tau\delta$, $|A_{21}z| - \mu < -\dfrac{3}{4}\delta$, and $a_{22}\tau < \dfrac{1}{4}$, we have

$$\dot{V} < |y| \left[2h\left(-\frac{3}{4}\delta + a_{22}\tau\delta\right) + K(1 + |y|)\right] \,.$$

Hence

$$\dot{V} < |y| \,(-\delta h + K(1 + \tau\delta)) \,.$$

Therefore if $h > K(1 + \tau\delta)/\delta$, in addition to the previous requirement that $h \geq \left[\dfrac{\mu - \delta}{\tau\delta}\right]^2 + \beta^2$, we have $\dot{V} < 0$ in $\Gamma \backslash \Omega$.

Let

$$C_1 = \frac{\mu^2}{\tau^2} \,, \quad C_2 = K(1 + \tau\mu) \,.$$

Note that τ depends only on β and a_{22}. Then $\dot{V} < 0$ in $\Gamma \backslash \Omega$ if $h \geq \max\left\{\dfrac{C_1}{\delta^2} + \beta^2 \,, \ \dfrac{C_2}{\delta}\right\}$ and the proof is complete.

$\square$

In the following Sections we will extend the obtained results to multi-input systems. As in the case of single neuron type controllers we shall utilize ideas from the variable structure control. To proceed with the analysis we will need a method for designing a switching surface (hyperplane) for multi-input dynamic systems. This is the subject of the next Section.

It is important to observe that by using the variable structure systems approach we are able to circumvent the problems which arise from discontinuity of the nonlinearities which characterize neurons.

8. DESIGN OF THE SWITCHING HYPERPLANE FOR MULTI-INPUT SYSTEMS

In this Section we will briefly discuss a method for designing of the switching surface for multi-input systems. The method is based on that of El-Ghezawi et al. [8]. Certain relations which come out during the analysis of this method are instrumental in the construction of the state transformation discussed in the following Section.

Consider the equivalent system

$$\dot{x} = [I_n - B(SB)^{-1}S]Ax .$$

It is easy to see that $B(SB)^{-1}S$ is a projector and has rank m. Hence $I_n - B(SB)^{-1}S$ is also a projector with rank n$-$m. Therefore the matrix $A_{eq} = [I_n - B(SB)^{-1}S]A$ in the equivalent system can have at most n$-$m nonzero eigenvalues. Our goal is to choose S so that the nonzero eigenvalues of A_{eq} are prescribed negative real numbers and the corresponding eigenvectors $\{w_1,...,w_{n-m}\}$ are to be chosen. Let $W = [w_1...w_{n-m}]$; note that $W \in \mathbb{R}^{n \times (n-m)}$. In sliding mode, the system is described by

$$\dot{x} = A_{eq}x$$
$$\sigma(x) = Sx = 0 .$$

The order of the system is n$-$m and the solution must be in the null space of S, that is, SW = 0. It is well known that complete controllability of the pair (A,B) is equivalent to the existence of a controller of the form $u = -Kx$ so that the eigenvalues of $A - BK$ can be arbitrarily assigned [6]. Our equivalent system has the form

$$\dot{x} = Ax - B[(SB)^{-1}SA]x .$$

If we let $K = (SB)^{-1}SA$, we need $A - BK$ to have n$-$m prescribed negative eigenvalues $\{-\lambda_1,...,-\lambda_{n-m}\}$ and n$-$m corresponding eigenvectors $\{w_1,...,w_{n-m}\}$. This is equivalent to

$$(A - BK)W = WJ \tag{8.1}$$

where $J = \text{diag}[-\lambda_1,...,-\lambda_{n-m}]$.

Denote by R(T) the range of the operator T. Since we requires SB to be nonsingular and SW = 0, we must have

$$R(B) \cap R(W) = \{0\} . \tag{8.2}$$

It then follows that we should choose the generalized inverses B^g, W^g of B, W so that

$$B^g W = 0 \tag{8.3a}$$

and

$$W^g B = 0 . \tag{8.3b}$$

We choose $\{w_1,...,w_{n-m}\}$ so that (8.3b) holds. We can now construct S. Let $W^\perp \in \mathbb{R}^{m \times n}$ be any full rank annihalator of W, that is $W^\perp W = 0$. Since a necessary condition for Sx = 0 to be a switching surface is SW = 0, we see that $QW^\perp$, for any nonsingular $Q \in \mathbb{R}^{m \times m}$, is a candidate. We also require that $SB = I_m$. Note that since $R(W) \cap R(B) = \{0\}$, $W^\perp B$ is invertible. We let $Q = (W^\perp B)^{-1}$ and let $S = QW^\perp$. It is easy to see that $SB = I_m$ and hence $(W^\perp B)^{-1}W^\perp$ is a generalized inverse of B. If we let $B^g = S$ in (8.3a), then the condition is satisfied.

We will utilize the results of this Section to construct a state-space transformation decoupling the neural type controller from the rest of the system.

9. DECOUPLING THE NEURAL CONTROLLER FROM THE REST OF THE SYSTEM

In this Section we introduce a transformation which brings the closed-loop system into the new coordinates in which the neural structure of the controller is revealed. This transformation will also facilitate the task of estimating stability regions. The results of this Section are based on the paper by Madani-Esfahani et al. [19].

Let $M \in \mathbb{R}^{n \times n}$ be defined by

$$M = \begin{bmatrix} W^g \\ S \end{bmatrix},$$

where W^g is defined by (8.3b). Note that M is invertible with $M^{-1} = [W \ B]$. Introduce the new coordinates

$$\hat{x} = Mx .$$

Let $z = W^g x$ and $y = Sx$. Then $\hat{x} = \begin{bmatrix} z \\ y \end{bmatrix}$. In the new coordinates, the system becomes

$$\hat{x} = MAM^{-1}\hat{x} + MBu .$$

We write

$$MAM^{-1} = \begin{bmatrix} A_{11} & A_{12} \\ A_{21} & A_{22} \end{bmatrix}$$

where $A_{11} \in \mathbb{R}^{(n-m) \times (n-m)}$, $A_{22} \in \mathbb{R}^{m \times m}$. Note that

$$MB = \begin{bmatrix} 0 \\ I_m \end{bmatrix} .$$

hence

$$\begin{aligned} \dot{z} &= A_{11}z + A_{12}y \\ \dot{y} &= A_{21}z + A_{22}y + u . \end{aligned} \tag{9.1}$$

Observe that $y = \sigma$ and that $A_{21}z + A_{22}y = SAM^{-1}\hat{x} = SAx = -u_{eq}$. Thus (9.1) can be rewritten as

$$\begin{cases} \dot{z} = A_{11}z + A_{12}\sigma \\ \dot{\sigma} = -u_{eq} + u . \end{cases} \tag{9.2}$$

From (8.1) we have

$$(A - BK)W = WJ .$$

Hence

$$W^g AW = J$$

since $W^g B = 0$ and $W^g W = I_{n-m}$. We know that $A_{11} = W^g AW$, and therefore $A_{11} = J$.

Example 9.1. Consider the following system:

$$\dot{x} = \begin{bmatrix} 0 & 1 \\ 3 & 2 \end{bmatrix} x + \begin{bmatrix} 0 \\ 1 \end{bmatrix} u .$$

We would like to design the switching surface so that the system restricted to this surface is stable and has eigenvalue -1. Suppose we choose the corresponding eigenvector for the equivalent system to be $W = w_1 = \begin{bmatrix} 1 \\ -1 \end{bmatrix}$. One can easily check that

$$W^g = [1 \ \ 0] \ , \ \ B^g = S = [1 \ \ 1]$$

satisfy (8.3). Hence the switching surface is $\sigma(x) = x_1 + x_2 = 0$. The transformation matrix is

$$M = \begin{bmatrix} W^g \\ S \end{bmatrix} = \begin{bmatrix} 1 & 0 \\ 1 & 1 \end{bmatrix}.$$

The system in the new coordinates is:

$$\dot{z} = -z + \sigma$$
$$\dot{\sigma} = 3\sigma + u \ .$$

The system restricted to the switching surface is governed by

$$\dot{z} = -z \ .$$

$\square$

In what follows we analyze the closed-looped system (9.1) with the controller

$$u = - \begin{bmatrix} \mu_1 \operatorname{sgn}\sigma_1 \\ \vdots \\ \mu_m \operatorname{sgn}\sigma_m \end{bmatrix}. \tag{9.3}$$

For convenience, we let

$$D = \operatorname{diag}[\mu_1, ..., \mu_m]$$

and

$$\operatorname{sgn}\sigma = \begin{bmatrix} \operatorname{sgn}\sigma_1 \\ \vdots \\ \operatorname{sgn}\sigma_m \end{bmatrix}.$$

We can now write (9.3) as

$$u = -D \operatorname{sgn}\sigma \ . \tag{9.4}$$

Combining (9.1) and (9.4) yields

$$\left. \begin{aligned} \dot{z} &= A_{11}z + A_{12}y \\ \dot{y} &= A_{21}z + A_{22}y - D \operatorname{sgn} \sigma \end{aligned} \right\} \tag{9.5}$$

Remark 9.1.

Note that the subsystem

$$\dot{y} = A_{21}z + A_{22}y - D \text{ sgn } \sigma$$

which can be interpreted as a dynamic controller driving the dynamic system

$$\dot{z} = A_{11}z + A_{12}y$$

has a structure of an additive neural network model. Although we arrived at (9.5) starting with the controller (9.4) whose structure does not correspond to an additive neural network model, we can utilize the above analysis in the case when we explicitly apply a neural control strategy. We proceed as follows. Suppose we are given a dynamic system model

$$x = Ax + Bu$$

We apply an additive neural network control law

$$\dot{u} = \beta u - D \text{ sgn } \sigma(x,u) + C^T x , \tag{9.6}$$

where $\beta \in \mathbb{R}^{m \times m}$. The closed-loop system is

$$\begin{bmatrix} \dot{x} \\ \dot{u} \end{bmatrix} = \begin{bmatrix} A & B \\ C^T & \beta \end{bmatrix} \begin{bmatrix} x \\ u \end{bmatrix} + \begin{bmatrix} 0 \\ I_m \end{bmatrix} (-D \text{ sgn } \sigma(x,u)), \tag{9.7}$$

where $[x^T, u^T]^T \in \mathbb{R}^{n+m}$, and $\sigma(x,u)$ is a switching surface to be chosen. Using the approach presented in Section 8 we design the switching hyperplane $\sigma(x,u)$ and then construct the transformation M following the development in Section 9. In the new coordinates (9.7) will have the form (9.5), where now $A_{11} \in \mathbb{R}^{n \times n}$, $A_{22} \in \mathbb{R}^{m \times m}$, $z \in \mathbb{R}^n$ and $y \in \mathbb{R}^m$. We know that the above procedure yields a stable closed-loop system (see Section 4). However, we are also interested in the extent of the stability properties of the closed-loop system. The next Section deals with this issue.

10. ESTIMATION OF STABILITY REGIONS OF DYNAMIC SYSTEMS DRIVEN BY THE NEURAL CONTROLLERS

This Section is devoted to the problem of estimating sliding domains of a class of systems modeled by (9.5). The development of this Section follows closely the arguments of Madani-Esfahani et al. [19]. In the analysis we shall use the following notation.

For i = 1,2, j = 1,2, we let

$$a_{ij} = ||A_{ij}|| \ ,$$

where the A_{ij}'s are from (9.5) and $||A_{ij}|| = \max_{x}\{||A_{ij}x||_2 \mid ||x||_2 \le 1\}$.

Let

$$\mu = \min\{\mu_1,...,\mu_m\} \ ,$$
$$\lambda = \min\{|\lambda_1|,..., |\lambda_{n-m}|\} \ .$$

Note that the controllability of the pair (A,B) implies that A_{12} is not zero and hence $a_{12} \ne 0$.

We will consider the cases A_{21} is 0 and A_{21} not necessarily 0 separately in Subsections 10.1 and 10.2. The case of $A_{21} = 0$ is simpler and gives a flavor of the argument used in the general case. We obtain explicit bounds on the time it takes to reach the switching surface for both cases. Here again, as in the single-input case, we utilize a variable structure approach. This will guard us against problems caused by the discontinuous nature of the nonlinearities which characterize neural network models.

10.1. $A_{21} = 0$.

We need the following lemma.

Lemma 10.1. Suppose $\phi(t)$ is real-valued and $k \ne 0$. If

$$\dot{\phi} - k\,\phi \le -\mu, \quad \mu \in \mathbb{R} \ ,$$

then for $t \ge t_0$,

$$\phi(t) \le \frac{\mu}{k} + (\phi(t_0) - \frac{\mu}{k})\,e^{k(t-t_0)} \ .$$

Proof. Note that $\dot{\phi} - k\,\phi \le -\mu$ is equivalent to $\frac{d}{dt}(e^{-kt}\phi) \le -\mu e^{-kt}$. The conclusion is obtained by integration. $\qquad\square$

The system to be considered in this Subsection has the form:

$$\dot{z} = A_{11}z + A_{12}\sigma$$
$$\dot{\sigma} = A_{22}\sigma - D\text{sgn}\sigma. \tag{10.1}$$

Suppose $\sigma(t_0) = \sigma_0 \in \mathbb{R}^m$. If $\sigma \neq 0$, then

$$\frac{d\|\sigma\|}{dt} = \frac{\sigma^T \dot{\sigma}}{\|\sigma\|} = \sigma^T A_{22} \frac{\sigma}{\|\sigma\|} - \frac{1}{\|\sigma\|} \sum_{i=1}^{m} \mu_i |\sigma_i|.$$

Since $\sum_{i=1}^{m} |\sigma_i| \geq \|\sigma\|$, we have

$$\frac{d\|\sigma\|}{dt} \leq \|A_{22}\| \|\sigma\| - \mu = a_{22}\|\sigma\| - \mu.$$

Note that if $0 < \|\sigma\| < \dfrac{\mu}{a_{22}}$, then $\dfrac{d\|\sigma\|}{dt} < 0$. Hence $\sigma(t_1) = 0$ implies that $\sigma(t) = 0$

for $t \geq t_1$. Also $\dfrac{d\|\sigma\|}{dt} < 0$ is equivalent to the condition $\sigma^T \dot{\sigma} < 0$ we have in Section 4. By Lemma 10.1,

$$\|\sigma(t)\| \leq \frac{\mu}{a_{22}} + (\|\sigma_0\| - \frac{\mu}{a_{22}}) e^{a_{22}(t-t_0)}.$$

Therefore if $\|\sigma_0\| < \dfrac{\mu}{a_{22}}$, then $\sigma(t)$ is 0 for some finite t with

$$t \leq t_0 + \frac{1}{a_{22}} \left[\log\mu - \log(\mu - a_{22}\|\sigma_0\|) \right]. \tag{10.2}$$

Observe that if A_{22} is stable, then $\mathbb{R}^n$ is a region of asymptotic stability and the switching surface is reached in finite time. Otherwise, a region of asymptotic stability is given by $\{x \in \mathbb{R}^n : \|Sx\| < \dfrac{\mu}{a_{22}}\}$. In both cases, $\sigma = 0$ is a sliding domain.

Example 9.1. (continued). Let $u = -10 \text{ sgn } \sigma$. The closed-loop system in the new coordinates has the form:

$$\dot{z} = -z + \sigma$$
$$\dot{\sigma} = 3\sigma - 10 \text{ sgn}\sigma.$$

In this case, $A_{22} = 3$ is unstable. The switching surface is $\sigma(x) = x_1 + x_2 = 0$. From the above, a region of asymptotic stability is given by:

$$\mathscr{R} = \{(x_1, x_2) \mid |x_1 + x_2| < \frac{10}{3}\}.$$

Actually $\mathscr{R}$ is the region of asymptotic stability (see Fig. 10).

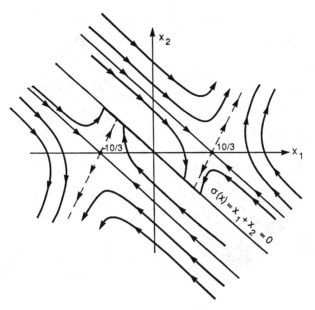

Fig. 10. Phase-plane portrait for Example 9.1.

10.2. The general case when A_{21} is not necessarily zero

We now have

$$\dot{z} = A_{11}z + A_{12}\sigma$$

$$\dot{\sigma} = A_{21}z + A_{22}\sigma - \mathrm{Dsgn}\sigma.$$

Suppose $\|z\| \neq 0$, $\|\sigma\| \neq 0$. Then

$$\frac{d\|\sigma\|}{dt} = \frac{\sigma^T \dot{\sigma}}{\|\sigma\|} = \frac{\sigma^T}{\|\sigma\|} A_{21}z + \sigma^T A_{22} \frac{\sigma}{\|\sigma\|} - \sum_{i=1}^{m} \mu_i \frac{|\sigma_i|}{\|\sigma\|},$$

and

$$\frac{d\|z\|}{dt} = \frac{z^T \dot{z}}{\|z\|} = z^T A_{11} \frac{z}{\|z\|} + \frac{z^T}{\|z\|} A_{12} \sigma \, .$$

After some manipulations we obtain

$$\frac{d\|\sigma\|}{dt} \le a_{21} \|z\| + a_{22} \|\sigma\| - \mu \tag{10.3}$$

and

$$\frac{d\|z\|}{dt} \le -\lambda \|z\| + a_{12} \|\sigma\| \, . \tag{10.4}$$

Let

$$\Sigma_1 = \left\{ (z, \sigma) \mid \|\sigma\| < \frac{\lambda \mu}{a_{12} a_{22} + \lambda a_{22}} \, , \quad z \in \mathbb{R}^{n-m} \, , \quad \sigma \in \mathbb{R}^m \right\} \, ,$$

$$\Sigma_2 = \left\{ (z, \sigma) \mid a_{21} \|z\| + a_{22} \|\sigma\| < \mu \, , \quad z \in \mathbb{R}^{n-m} \, , \quad \sigma \in \mathbb{R}^m \right\} \, ,$$

and

$$\Sigma = \Sigma_1 \cap \Sigma_2 \, .$$

Theorem 10.2. A trajectory that starts in Σ stays in Σ and reaches the switching surface in finite time, which implies that Σ is a region of asymptotic stability (RAS).

Proof. Let

$$N_1 = \{ (z, \sigma) \mid a_{12} \|\sigma\| \ge \lambda \|z\| \}$$

and

$$N_2 = \{ (z, \sigma) \mid a_{12} \|\sigma\| < \lambda \|z\| \}$$

(see Fig. 11). By (10.3) we have $\dfrac{d\|\sigma\|}{dt} < 0$ in Σ. Hence if $(z(t_0), \sigma(t_0)) \in \Sigma$, we have $(z(t), \sigma(t)) \in \Sigma_1$ for $t \ge t_0$. For $(z(t), \sigma(t)) \in N_1 \cap \Sigma \subset N_1 \cap \Sigma_1$, we have

$$a_{21} \|z\| + a_{22} \|\sigma\|$$
$$\le \frac{a_{21} a_{12}}{\lambda} \|\sigma\| + a_{22} \|\sigma\|$$
$$\le \frac{a_{22} a_{12}}{\lambda} \|\sigma(t_0)\| + a_{22} \|\sigma(t_0)\|$$
$$< \mu - \epsilon \quad \text{for some } \epsilon > 0 \, .$$

Thus we can conclude that a trajectory $(z(t), \sigma(t))$ can leave Σ only through $N_2 \cap \Sigma$.

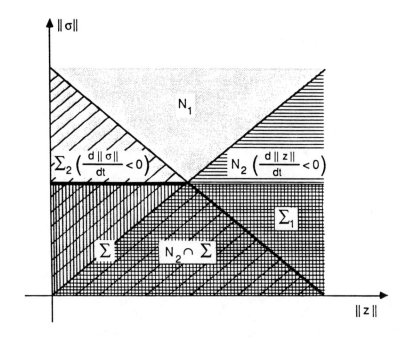

Fig. 11. Illustration of regions used in the proof of Theorem 10.2.

However we have $\dfrac{d\|\sigma\|}{dt} < 0$ in Σ and $\dfrac{d\|z\|}{dt} < 0$ in N_2, and hence $a_{21}\|z\| + a_{22}\|\sigma\|$ is a decreasing function in $N_2 \bigcap \Sigma$. Therefore a trajectory cannot leave $N_2 \bigcap \Sigma$. Hence a trajectory which starts in Σ stays in Σ.

Suppose $(z(t),\ \sigma(t)) \in N_1 \bigcap \Sigma$ for $t_1 \le t \le t_2$. Then we have

$$\frac{d\|\sigma\|}{dt} \le a_{21}\|z\| + a_{22}\|\sigma\| - \mu$$

$$\le \left(\frac{a_{12}a_{21}}{\lambda} + a_{22} \right) \|\sigma\| - \mu \,.$$

Let $k = \dfrac{a_{12}a_{21}}{\lambda} + a_{22}$. Then by Lemma 10.1, we have

$$\|\sigma(t_2)\| \leq \|\sigma(t_1)\| e^{k(t_2-t_1)} - \frac{\mu}{k}\left\{e^{k(t_2-t_1)} - 1\right\}.$$

Therefore

$$\|\sigma(t_1)\| - \|\sigma(t_2)\| \geq \left[\frac{\mu}{k} - \|\sigma(t_1)\|\right]\left(e^{k(t_2-t_1)} - 1\right)$$

$$\geq \left[\frac{\mu}{k} - \|\sigma(t_1)\|\right]k(t_2 - t_1)$$

$$= (\mu - k\|\sigma(t_1)\|)(t_2 - t_1).$$

Suppose $\sigma(t_0) = \sigma_0$.

Since $\|\sigma(t)\|$ is decreasing in Σ, we have

$$\|\sigma(t_1)\| - \|\sigma(t_2)\| \geq (\mu - k\|\sigma_0\|)(t_2 - t_1).$$

We conclude that a trajectory cannot spend an infinite amount of time in $N_1 \cap \Sigma$ with $\|\sigma(t)\| > 0$.

We claim that if $(z(t_1), \sigma(t_1)) \in N_2 \cap \Sigma$, $(z(t_2), \sigma(t_2)) \in N_2 \cap \Sigma$, and $t_1 < t_2$, then $\|z(t_2)\| < \|z(t_1)\|$. The claim is clear if $(z(t), \sigma(t)) \in N_2 \cap \Sigma$ for $t_1 \leq t \leq t_2$. Otherwise, suppose $(z(t), \sigma(t)) \in N_2 \cap \Sigma$ for $t_1 < t \leq T_1$, $T_2 \leq t < t_2$, and $(z(T_1), \sigma(T_1))$, $(z(T_2), \sigma(T_2))$ are on the boundary of $N_2 \cap \Sigma$. Therefore

$$\frac{\|z(T_1)\|}{\|\sigma(T_1)\|} = \frac{\|z(T_2)\|}{\|\sigma(T_2)\|}.$$

Since $\|\sigma(t)\|$ decreases in Σ, we have $\|z(T_2)\| < \|z(T_1)\|$ and we can conclude that $\|z(t_2)\| < \|z(t_1)\|$ as in the case where the whole segment is in $N_2 \cap \Sigma$.

Hence if $(z(t_0), \sigma(t_0)) \in N_2 \cap \Sigma$, we have for $t \geq t_0$,

$$\frac{d\|\sigma\|}{dt} \leq a_{21}\|z(t_0)\| + a_{22}\|\sigma(t_0)\| - \mu < 0.$$

Therefore, if $(z(t), \sigma(t)) \in N_2 \cap \Sigma$ for $t_1 \leq t \leq t_2$ then

$$\|\sigma(t_1)\| - \|\sigma(t_2)\| \geq (\mu - a_{21}\|z(t_0)\| - a_{22}\|\sigma(t_0)\|)(t_2 - t_1).$$

We can conclude that a trajectory cannot spend an infinite amount of time in $N_2 \cap \Sigma$ with $\sigma(t) \neq 0$. Thus we must reach the switching surface $\sigma = 0$ in finite time if we start in Σ.

$\square$

From the above, we can give explicit estimates of the time it takes to reach the switching surface starting in Σ.

Corollary 10.3. Let

$$\beta = \min\left\{\mu - (\frac{a_{12}a_{21}}{\lambda} + a_{22})\|\sigma(t_0)\|, \ \mu - a_{21}\|z(t_0)\| - \|\sigma(t_0)\|\right\}$$

Starting at $(z(t_0), \sigma(t_0)) \in \Sigma$, we must reach $\sigma = 0$ in

$$t \leq t_0 + \frac{\|\sigma(t_0)\|}{\beta} .$$

Example 10.1. Consider the following system:

$$\dot{x} = \begin{bmatrix} 3 & 1 & 1 \\ -6 & 1 & 0 \\ 0 & 0 & 3 \end{bmatrix} x + \begin{bmatrix} 0 & 0 \\ 1 & 1 \\ 0 & 1 \end{bmatrix} u$$

with $|u_i| \leq 10$. Suppose the desired eigenvalue of the reduced order system is -3 with corresponding eigenvector $w_1 = \begin{bmatrix} 1 \\ -6 \\ 0 \end{bmatrix}$. One can check that

$$W^g = w_1^g = [1 \ 0 \ 0], \quad S = B^g = \begin{bmatrix} 6 & 1 & -1 \\ 0 & 0 & 1 \end{bmatrix}$$

satisfy (8.3). The transformation matrix is

$$M = \begin{bmatrix} 1 & 0 & 0 \\ 6 & 1 & -1 \\ 0 & 0 & 1 \end{bmatrix} .$$

The system in the new coordinates has the form:

$$\dot{z} = -3z + [1 \ 2] \sigma$$

$$\dot{\sigma} = \begin{bmatrix} -30 \\ 0 \end{bmatrix} z + \begin{bmatrix} 7 & 10 \\ 0 & 3 \end{bmatrix} \sigma - 10 \, \text{sgn} \, \sigma .$$

In this case, $\mu = 10$, $\lambda = 3$, $a_{12} = 2.24$, $a_{21} = 30$, $a_{22} = 12.46$. Hence

$$\Sigma = \{(z,y) : \|y\| < \frac{30}{104.58}, \ 30\|z\| + 12.46\|y\| < 10\} .$$

$\square$

Using a Lyapunov function argument, we can give another region of asymptotic stability.

Theorem 10.4. A region of asymptotic stability of the system is

$$\mathcal{R} = \{(z,\sigma) \mid \alpha \|z\| + \beta \|\sigma\| < \mu\}$$

where $\beta = \dfrac{1}{2}\left[(a_{22}-\lambda) + \sqrt{(a_{22}+\lambda)^2 + 4a_{12}a_{21}}\right]$ and $\alpha = \beta a_{21}/(\beta+\lambda)$.

Proof. Let V be the positive definite function defined by

$$V(z,\sigma) = \alpha \|z\| + \beta \|\sigma\| .$$

The Lyapunov derivative is

$$\dot{V}(z,\sigma) = \alpha \frac{z^T \dot{z}}{\|z\|} + \beta \frac{\sigma^T \dot{\sigma}}{\|\sigma\|} .$$

From (10.3), (10.4) we get

$$\dot{V}(z,\sigma) \leq \alpha(-\lambda\|z\| + a_{12}\|\sigma\|) + \beta(a_{21}\|z\| + a_{22}\|\sigma\| - \mu)$$
$$= (-\lambda\alpha + \beta a_{21})\|z\| + (\alpha a_{12} + a_{22}\beta)\|\sigma\| - \mu\beta .$$

Using the values of α and β, we have

$$\dot{V}(z,\sigma) \leq \beta[\alpha\|z\| + \beta\|\sigma\| - \mu]$$

The right hand side is less than 0 in $\mathcal{R}$. This finishes the proof if $z(t)\neq 0$ and $\sigma(t)\neq 0$.

Otherwise observe that $\beta \geq a_{22}$ and $\alpha \leq a_{21}$. By (10.3) and (10.4) we have on $\mathcal{R} \cap \{z = 0\}$

$$\frac{d\|\sigma\|}{dt} \leq a_{22}\|\sigma\| - \mu \leq \beta\|\sigma\| - \mu < 0$$

and on $\mathcal{R} \cap \{\sigma = 0\}$

$$\frac{d\|z\|}{dt} \leq -\lambda\|z\| < 0 .$$

Hence $V(z,\sigma)$ decreases on the critical surfaces also and we are done.

$\square$

As a consequence of Theorems 10.2 and 10.4 we give a new region of asymptotic stability with sliding.

Theorem 10.5. Let

$$\mathcal{R}_1 = \{(z,\sigma) \,|\, \alpha\|z\| + \beta\|\sigma\| < \mu, \, \|\sigma\| \geq \frac{\lambda}{a_{12}}\|z\|\}$$

$$\cup \, \{(z,\sigma) \,|\, a_{21}\|z\| + a_{22}\|\sigma\| < \mu, \, \|\sigma\| \leq \frac{\lambda}{a_{12}}\|z\|\} \, .$$

Then $\mathcal{R}_1$ is a region of asymptotic stability with sliding.

Proof. We use the same notation as in Theorems 10.2 and 10.4. Observe that

$$\mathcal{R}_1 = \Sigma \cup (\mathcal{R} \cap N_1) \, .$$

For a trajectory that starts in $\mathcal{R} \cap N_1$ to reach the switching surface $\{\sigma=0\}$, it must pass through Σ, which is a region of asymptotic stability with sliding. See Fig. 12.

□

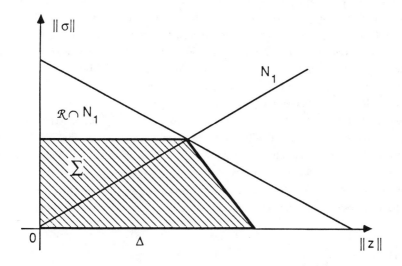

Fig. 12. Illustration of the proof of Theorem 10.5.

Example 10.1. (continued). We have found

$$\Sigma = \{(z,\sigma) \,|30\|z\| + 12.46\|\sigma\| < 10, \; \|\sigma\| < 0.287\} \,.$$

We have $\alpha = 25.263$ and $\beta = 15.997$. Therefore

$$\mathscr{R} = \{(z,\sigma) \,|25.263\|z\| + 15.997\|\sigma\| < 10\}$$

and

$$\mathscr{R}_1 = \Sigma \cup (\mathscr{R} \cap N_1) \,.$$

Recall that $N_1 = \{(z,\sigma) \,|\|\sigma\| > 1.34\|z\|\}$.

11. CONCLUDING REMARKS

In this chapter we investigated viability of employing controllers based on additive neural network models to the problem of stabilization (tracking) of a class of dynamic systems. Two approaches to designing stabilizing controllers were proposed. Elements of the variable structure control theory were utilized to construct such controllers. The proposed controllers are characterized by robustness property which is inherent in the variable structure controllers. An important role in the analysis was played by a special state space transformation. This transformation not only facilitated the stability analysis but also helped to utilize additive neural network models in designing stabilizing controllers. The proposed approach is promising in three ways. First, it results in robust controllers. Second, it has a potential to be employed in constructing fault tolerant controllers. Third, it allowed us to circumvent stability analysis problems caused by the discontinuous nonlinearities which describe neurons. Also generalizations to the control of a more general class of dynamic systems are feasible. The proposed approach in this chapter and the results of Walcott and Żak [27] constitute a nice starting point to designing neural network based state estimators for dynamic systems.

REFERENCES

[1]   J. S. Albus, "*A new approach to manipulator control: The Cerebellar Model Articulation Controller (CMAC),*" Trans. of the ASME, J. Dynamic Systems, Measurement, and Control, Vol. 97, Series G, No. 3, pp. 220-227, Sept. 1975.

[2] J. S. Albus, *"Data storage in the Cerebellar Model Articulation Controller (CMAC),"* Trans. of the ASME, J. Dynamic Systems, Measurement, and Control, Vol. 97, Series G, No. 3, pp. 228-233, Sept. 1975.

[3] L. O. Chua and G.-N. Lin, *"Nonlinear programming without computation,"* IEEE Trans. Circuits Syst., Vol. CAS-31, No. 2, pp. 182-188, Feb. 1984.

[4] L. O. Chua and L. Yang, *"Cellular neural networks: Theory,"* IEEE Trans. Circuits Syst., Vol. 35, No. 10, pp. 1257-1272, Oct. 1988.

[5] E. J. Davison, *"The output control of linear time-invariant multivariable systems with unmeasurable arbitrary disturbances,"* IEEE Trans. Automat. Contr., Vol. AC-17, No. 5, pp. 621-630, Oct. 1972.

[6] R. A. DeCarlo, *"Linear Systems: A State Variable Approach with Numerical Implementation,"* Prentice Hall, Englewood Cliffs, New Jersey, 1989.

[7] R. A. DeCarlo, S. H. Żak, and G. P. Matthews, *"Variable structure control of nonlinear multivariable systems: A tutorial,"* Proceedings of the IEEE, Vol. 76, No. 3, pp. 212-232, March 1988.

[8] O. M. E. El-Ghezawi, A. S. I. Zinober, and S. A. Billings, *"Analysis and design of variable structure systems using a geometric approach,"* Int. J. Control, Vol. 38, No. 3, pp. 657-671, 1983.

[9] G. F. Franklin, J. D. Powell, and A. Emani-Naeini, *"Feedback Control of Dynamic Systems,"* Addison-Wesley, Reading, Massachusetts, 1986.

[10] H. J. Greenberg, *"Equilibria of the Brain-State-in-a-Box (BSB) neural model,"* Neural Networks, Vol. 1, No. 4, pp. 323-324, 1988.

[11] S. Grossberg, *"Nonlinear neural networks: Principles, mechanisms, and architectures,"* Neural Networks, Vol. 1, No. 1, pp. 17-61, 1988.

[12] M. Hached, S. M. Madani-Esfahani, and S. H. Żak, *"Stabilization of uncertain systems subject to hard bounds on control with application to a robot manipulator,"* IEEE J. Robotics and Automation, Vol. 4, No. 3, pp. 310-323, June 1988.

[13] J. J. Hopfield, *"Neural networks and physical systems with emergent collective computational abilities,"* Proc. Natl. Acad. Sci. USA, Vol. 79, pp. 2554-2558, April 1982.

[14] J. J. Hopfield, *"Neurons with graded response have collective computational properties like those of two-state neurons,"* Proc. Natl. Acad. Sci. USA, Vol. 81, pp. 3088-3092, May 1984.

[15] M. Kawato, Y. Uno, M. Isobe, and R. Suzuki, *"Hierarchical neural network model for voluntary movement with application to robotics,"* IEEE Control Systems Magazine, Vol. 8, No. 2, pp. 8-16, April 1988.

[16] M. P. Kennedy and L. O. Chua, *"Neural networks for nonlinear programming,"* IEEE Trans. Circuits Syst., Vol. 35, No. 5, pp. 554-562, May 1988.

[17] J-H. Li, A. N. Michel, and W. Porod, *"Analysis and synthesis of a class of neural networks: Variable structure systems with infinite gain,"* IEEE Trans. Circuits and Systems, Vol. 36, No. 5, pp. 713-731, May 1989.

[18] R. P. Lippmann, *"An introduction to computing with neural nets,"* IEEE ASSP Magazine, Vol. 4, No. 2, pp. 4-22, April 1987.

[19] S. M. Madani-Esfahani, S. Hui, and S. H. Żak, *"On the estimation of sliding domains and stability regions of variable structure control systems,"* Proc. 26th Annual Allerton Conf. Communication, Control, and Computing, Monticello, Illinois, pp. 518-527, Sept. 28-30, 1988.

[20] S. M. Madani-Esfahani, S. Hui, and S. H. Żak, *"Estimation of regions of asymptotic stability with sliding for relay-control systems,"* Proc. 28th IEEE Conf. Decision and Control, Tampa, Florida, Dec. 13-15, 1989.

[21] A. N. Michel, J. A. Farrell, and W. Porod, *"Qualitative analysis of neural networks,"* IEEE Trans. Circuits Syst., Vol. 36, No. 2, pp. 229-243, Feb. 1989.

[22] Y.-H. Pao, *"Adaptive Pattern Recognition and Neural Networks,"* Addison-Wesley, Reading, Massachusetts, 1989.

[23] M. J. S. Smith and C. L. Portmann, *"Practical design and analysis of simple "neural" optimization circuit,"* IEEE Trans. Circuits Syst., Vol. 36, No. 1, pp. 42-50, Jan. 1989.

[24] D. W. Tank and J. J. Hopfield, *"Simple "neural" optimization networks: An A/D converter, signal decision circuit, and a linear programming circuit,"* IEEE Trans. Circuits Syst., Vol. CAS-33, No. 5, pp. 533-541, May 1986.

[25] Ya. Z. Tsypkin, *"Relay Control Systems,"* Cambridge University Press, Cambridge, Great Britain, 1984.

[26] V. I. Utkin, *"Variable structure systems with sliding modes,"* IEEE Trans. Automat. Contr., Vol. AC-22, No. 2, pp. 212-222, April 1977.

[27] B. L. Walcott and S. H. Żak, *"Combined observer-controller synthesis for uncertain dynamical systems with applications,"* IEEE Trans. Syst. Man Cybernetics, Vol. 18,

No. 1, pp. 88-104, Jan./Feb. 1988.

[28] B. Widrow and R. Winter, *"Neural nets for adaptive filtering and adaptive pattern recognition,"* Computer, Vol. 21, No. 3, pp. 25-39, March 1988.

[29] K.-K. D. Young, *"Controller design for a manipulator using theory of variable structure systems,"* IEEE Trans. Systems, Man, and Cybernetics, Vol. SMC-8, No. 2, pp. 101-109, Feb. 1978.

6

CONTROL OF LOCOMOTION
IN HANDICAPPED HUMANS

Popovic B. Dejan

ABSTRACT

An skill-based expert system is designed for control of assisted gait in handicapped humans based on heuristics of locomotion. Studies on heuristics approach pointed out the novel model of cyclic movements. Sequences of singular events with a specific locomotion cycle are the basis for this new model. These events are called gait invariants. The method is explained using an externally controlled and powered artificial organ. The active above-knee prosthesis (AKP) with cybernetic actuator is discussed for multimode operation. Organisation of the knowledge system for such a control system is described. Advantages and perspectives of heuristic control approach are derived through clinical development of self-contained active AKP and comparison with other control methods for gait restoration of the amputee.

1. INTRODUCTION

Multivariable nonlinear nature of motor control in man and animal can not be easily mastered by numerical methods. The large diversity of sensory driven functional motions encountered in nature represents a serious challenge to control theory. To overcome this difficulty and arrive at relatively simple control solution, a skill based expert system approach to the control of locomotion was proposed. This approach was derived by observing heuristics of human motor acts and neurophysiological organization of movements. The advantages of heuristic control of locomotion was demonstrate using a man-machine dynamical system (Popovic et al., 89; Popovic et al., 90). An above knee active prostheses (AKP) and Hybrid assistive system (HAS) for gait restoration of paralyzed humans were used for preliminary tests of this approach. The presented material illustrates the potential of the skill based expert systems for the control of motion in biomedical robotics, but it is extended to grasping in industrial robotics (Tomovic et al., 87).

2. SKILL-BASED EXPERT SYSTEM FOR MOVEMENTS CONTROL

Control has the task to introduce an ordering relation into a set of options. Heuristics offers a wide range of procedures, numerical, non - numerical, intuitive to solve a goal oriented task. Consequently, heuristic problem should begin with goal definition. But, when dealing with heuristics of motor skills we can observe only the final solution without being able to prove rigorously what are the selection criteria which have produced the observed deterministic performance. The learning process by which a skill becomes controlled by reflex and automatic mechanisms is also not understood. By multidisciplinary efforts of life and engineering sciences it should be possible in the future to arrive at a much better understanding of heuristics of motor skills. Such attempts are currently taking place (Tomovic, 88).

Assumptions about heuristics of functional motions are helpful as a general guiding principle in the design of skill-based expert systems. At this stage of our understanding of motor control in man, they are, however, unable to generate deductive procedures by which reflex and automatic mechanisms may be transferred to the computer. In view of this fact, an inductive approach to the design of skill based control of locomotion will be described.

The aim of the proposed inductive procedure for skill based expert system control of locomotion in handicapped humans is twofold. In the first place, its role is to identify the invariant features of biped gait modes common to all individuals and, secondly, to represent them in the knowledge base in such a way that they can be used for control. For that purpose special experimental data are needed in addition to available descriptions of invariants pertaining to the gait of the handicapped human.

In order to explain the specific features of the control method used we will concentrate on some of the features of an Skill-Based Expert Systems (SBES) in general.

The first characteristic of this technique is that it deals mainly with non-numerical symbols. The second characteristic of programs is that they attack problems for which no general algorithm is known - that is, there is no known sequence of steps guaranteed to lead to the solution. Since there is no known algorithms, heuristic approach may be applied. The heuristic procedure consists in choosing a method of attack which seems promising, while keeping open the possibility of changing to another if the first seems not to be leading quickly to a solution.

One of the features of programs is the possibility of knowledge representation, that is, a correspondence between the external world and a symbolic representation. This knowledge can usually be studied and understood in what we may call human terms or skill.

Another criterion of programming is the ability to learn from mistakes, that is, to improve the performance by taking account of past errors. The computer is often valued for its property, unlike the human mind, of forgetting nothing; but it is precisely the ability to forget that gives man his ability to learn, by putting aside unimportant details and replacing individual facts by procedures.

The first applied formalism, called production rules control, belongs to period well before the advent of artificial intelligence (AI) (Post, 43, Chomsky, 57). A production rule is a situation - action couple, meaning that whenever a certain situation is encountered, given as the left side of the rule the action on the right side of the rule is performed. There is no a priori constraint on the form of the situation or of the action. A system based on production rules will usually have three components: 1) the rule base, consisting of the set of production rules, 2) one or more data structures containing the known facts relevant to the domain of interest, possibly also some useful definitions; these are often called facts bases, and 3) the interpreter of these facts and rules, which is the mechanism that decides which rule to apply and initiates the corresponding action. It is fundamental principle of rule-base programming that each rule is an independent item of knowledge, containing all the conditions required for an application. There is no mechanism anywhere else except in the rule itself that creates conditions which could prevent it from being applied. The second principle is that the rules are ignorant of one another; only the interpreter knows what is happening so far the rules are concerned. In a pure production rule system the rules are not ordered in any way and in principle any one can be activated at any moment. Because of this modularity such a system can be easily modified because the addition, deletion or modification of a rule does not affects the architectural structure of the program.

The knowledge representation in an skill based expert system differs from the one in production rule system. Only rarely can the knowledge concerning a particular field be out in terms of a single formalism. There are usually some items of a heuristic nature. There are procedural items in addition to heuristics. A third type is factual knowledge. Finally, we will mention that the knowledge built into an skill-based expert system must consist of "tricks of the trade" in addition to the type of knowledge which may be expressed explicitly.

The above terminology is included in this paper in order to clarify the method which was applied for the control of artificial organs.

3. SYMBOLIC PRESENTATION OF THE KNOWLEDGE FOR GAIT CONTROL

The skill based expert system for an assistive system consists of following blocks: data, inference engine, regular rule base, operating rule base, data base and mode rule base. The regular rule base has two subsets of rules, regular and hazard rules. Regular rules are couples situation - action" within a specific gait mode. The part of the regular rule base dealing with conflict situations is called hazard rule base. Hazard states are situations where the hardware cannot produce adequate motor response or situations which are not expected in normal gait

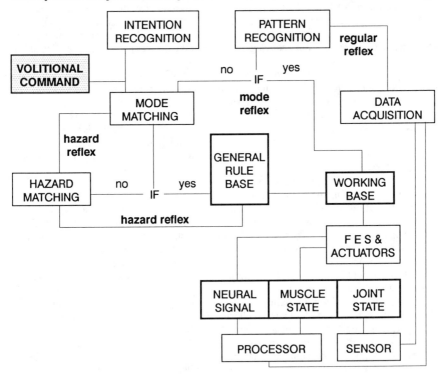

Fig 1. *Structural diagram of skill based expert system for an assistive system*

patterns. Term "firing the rule" is used for execution of an artificial reflex. Firing the rule from regular rule base results in a state of the neuroprosthetic actuator (locking, damping or extension - flexion of the joint). The operating rule base is the subset of the mode rule base or regular rule base. Firing of a rule from the mode base shifts the set of rules from the regular mode base to the operating base. The structure of such a knowledge base is shown in Fig 1. Rules pertaining to a gait mode are stored in an arbitrary way in the corresponding segments of the knowledge base. The transfer of the necessary rule block to the working base is governed by the so-called environmental recognizer whose duty is to relate environmental changes to the appropriate gait mode. Environment recognition is performed either by special rules, such as intention recognition, or by volitional intervention of the human.

Invariants for environment recognition (gait mode) and speed adaptation are difficult to estimate because kinematic and dynamic records of a "normal" gait for the paralyzed human are not known.

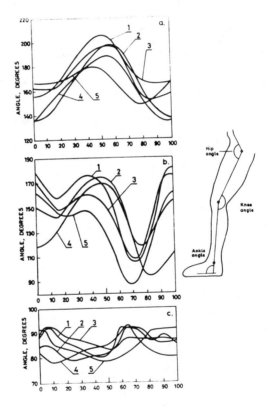

Fig 2. *Angular patterns for AKP. Velocity of the gait is $v_0 = 1.1$ m/s The AKP pattern is simulated with ankle cast eliminating ankle plantar flexion and limiting ankle dorsal flexion in normal subject. Dorsal flexion was limited to 8. a) Hip, b) Knee, c) ankle. 1 - level walking, 2 - up ramp, 3 - down ramp, 4 - up stairs, 5 - down stairs. Stairs walking was $v_0 = 0.6$ m/s*

In order to illustrate the technique the procedure of designing a rule will be used for a above-knee prosthesis. A record of the knee joint trajectory of an human while walking with an ankle cast is presented in Fig 2. The ankle cast is added in order to simulate the behaviour of the above-knee prosthesis which does not have an articulated ankle joint. Different types of ankle motion limitations were analyzed corresponding to different types of artificial foots. Such a record, the vertical attitude of the shank, and the sole pressure distribution are the minimal information needed for the identification of parameters involved in the multimode rule based control for an unilateral above-knee amputee. Foot pressure during locomotion (Soames, 85), hip and knee angle curves (Kirtley, 85, Berry, 52) are essential for the gait identification. Velocity, cadence, stride length and single support length can be adequately described by the foot pressure distribution, the knee angle curve and the shank displacement from the vertical line if AKP is in question (Popovic, 86). Analog information about joint motions cannot be used directly for the derivation of gait invariants. However, if the analog record is interpreted in terms of joint states (flexion, extension, loose, rigid), then the invariant cyclic succession of joint states for the required type of biped gait becomes transparent.The same applies to switching points related to state transitions.

In addition to analog records of joint trajectories of a given gait mode, the well known phase pattern of the human gait cycle, reproduced in Fig 3, is also needed for the knowledge representation of locomotion heuristics. Locomotion invariants displayed in Fig 3 are instrumental in matching sensory patterns to joint states. From Fig 3, it is easy to identify sensory input which is, in a unique way, related to the changes of joint states: heel and toe ground contacts, length of stance phase, magnitude of ground forces, passive and active terminal flexion and extension angles, displacement from the gravity direction (Nilsson,85). Sensory patterns associated with discrete changes of joint states are by this very fact discrete themselves. They consist of discrete values of variables representing the sensory part of control rules.

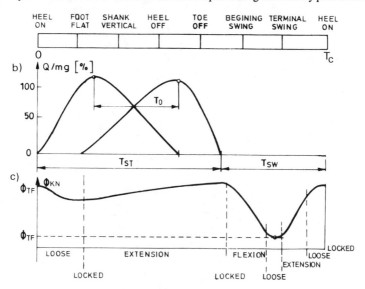

Fig 3. *Temporal patterns for gait assessment, T_C is a stride duration, T_{ST} is the stance phase length, T_{SW} the swing phase length,T_0 is the interval between maximal heel and toe ground reactions. b) normalized heel and toe zones ground reactions to the body weight in level walking, c)Knee angular pattern and knee joint state in level walking*

148

Reflex and automatic mechanisms involved in locomotion control can be formally represented in the knowledge base by production rules of the following type

$$B(x_1,x_2,x_3,...,x_n) \ -- \ C[y_1(k),y_2(k),... \ y_m(k)] \tag{1}$$

where $B(\cdot)$ is the boolean function; x_i are discrete sensory inputs; $C[\cdot]$ is the function of the four valued variable $y_j(k)$, $k =1,2,3,4$; which describes the state of the "j" joint at switching instants, m is the number of joints involved in locomotion. Above inductive approach to knowledge representation of gait heuristics has been successfully used to control AKP and Hybrid Assistive System (HAS) (Popovic, 88). After encouraging results with single mode rule based control of AKP, a more complex system for rehabilitation of lower limb amputees has been developed. Instead of single mode operation, the new system for the multimode control of the AKP offers several possibilities: walking on level ground, walking up and down the ramp and stairs, automatic speed adaptation, standing up and sitting down, walking backward and turning around (Popovic, et al., 90). The structure of such a knowledge base is shown in Fig 1. Applying the pattern matching procedure described by Expression (1), the set of control rules for each gait mode is thus derived.

4. MICROCOMPUTER REPRESENTATION OF REFLEX LIKE CONTROL

Reflex representation in microcomputer is in the form of a sequence of bytes describing the sensory input pattern which triggers the motor action. The number of bytes varies from reflex to reflex, depending on the number of inputs and actuators. Sequence length is also a function of the number of parameters and the complexity of the relevant boolean expression. In order to save memory, reflexes are memorized one next to each other. Current AKP controller is for bilateral use with up to eight analog inputs and two sets of switches. The sequence of bytes pertaining to a reflex reads as follows:

CL	CR	PL1	PL2	PR1	PR2	AL1	AL2	..	AL8	AR1	AR2	..	AR8	DL	DR	H	M

C byte defines the length and type of the reflex, L and R indexes are for the left (ipsilateral) and the right (contralateral) side, P for switch acquisition, A for analog input acquisition, D for actuator command, H and M for hazard and mode base activation. Details of byte organisation and assembler programming are described elsewhere (Popovic, 86, Tepavac 87).

Because of the resemblance of simple rules to biological reflexes a single rule in an skill based expert system is called artificial reflex. The artificial reflex activated in stable standing reads as follows:

$$[M*(H+T)=1] \ * \ [\alpha-(90-\delta\alpha)=1] \ * \ [\phi_K-(180-\delta\phi)=1] \ \Rightarrow \ B = 1$$

$\delta\alpha$ is a tolerance related to the vertical line, $\delta\phi$ is the tolerance for the knee angle. ϕ and α are first and second analog inputs, M, H and T are signs of switches at the heel, the mid-foot and the toe zone of the sole. The controlled system is one-joint with CA. Bytes describing this reflex are:

$$\{3C, F8, C2, 61, 53, C0, 61, C1, A8, 60, C0\}H = \{50\}H$$

The structure of the byte CL = {3C} H is:

b7	b6	b5	b4	b3	b2	b1	b0
0	1	1	0	1	1	0	0

b7 provides information of existence of the byte CR. If b7 =1, CR applies the same about the matching procedure of the contralateral leg. If b7 = 0, CR does not exists; b6 deals with the switch combination testing; b5 and b4 explore the switch combination length. Maximal length is (11)B. Last four bytes pertain to analog inputs.

PLi, i =1,2; These bytes define the switch configuration that satisfies the left side of a boolean expression. The significance of certain bits is as follows:

b7	b6	b5	b4	b3	b2	b1	b0
H	M	T		H	M	T	

H, M and T can be 0 or 1 and this corresponds to heel, mid-foot and toe zone switches at ipsilateral and contralateral sole. b4 and b0 deal with more combinations which satisfies the boolean expression. For example boolean expressions $M*(H+T)=1$ is satisfied (true) in the following cases:

H	M	T	$M*(H+T)$
0	1	1	1
1	1	0	1
1	0	1	1

Two bytes are needed for all combinations included by this expression. Their order is arbitrary. Bytes defining the permitted switch combinations have the following form:

H	M	T		H	M	T		H	M	T					
1	1	1	1	0	1	1	1	1	1	0	0	x	x	x	x

Sign x is used in columns which would not be tested.

ALk, k =1,3; The Alk defines the relation (limits) to be satisfied by the analog input. Bits b0, b1, b2 and b3 of the control byte C1 determine which analog input (k =1,2,3,4) takes part in the boolean expression defining the left side of the given rule.
They have the following form:

b7	b6	b5	b4	b3	b2	b1	b0
<	=	>		<	=	>	

Bits b7-b5 refer to the analog input, bits b3-b1 to first derivative of the same analog input. If b4 =1 it is necessary to test first derivative, if it is 0 the derivative should not be tested. b0 bit provides the information of existence of other bytes related to the same analog input.

ALk, k =2,4; They are appearing only together with ALk, k =1,3. These are parameters in hexadecimal form which are matched to the sensory input in the way described by A1k.

ARk*, k =1,2,3,4 are appearing only if corresponding Alk, k =1,2,3,4 has b0 =1 and it has the same form as ALk.

DL; This byte describes the action which has to be taken by the actuator. Bits of the byte D are:

b7	b6	b5	b4	b3	b2	b1	b0
x	y	x	y	x	y	x	y
1		2		3		4	

Combinations of bits x and y determine the actuator states:

actuators

x	y	driver output
0	0	0
0	1	-V
1	0	+V
1	1	0

In case the boolean expression, whose coincidence with the input vector is tested, doesn't define a reflex from the regular reflex base, but either one of hazard reflexes or any other reflex from the mode base, byte D becomes FF hexadecimal and thus defines no action. The actuator state doesn't change.

H and M; D byte value determines whether they exist or not. {FF}H means that the next two bytes (H and L) provide the address from which program execution is to be continued. This method enables inclusion of some special, additional, procedures as required by some of the rules. This increases the software system flexibility and leaves room for further system build-up and defining more complex system response to the occurrence of some (or all) characteristic inputs.

5. MULTIPLE JOINT CONTROL

An adaptive real time control for multiple joint system based on skill-based expert system uses hierarchical organization with following levels: 1) Local controller at each of actuators, 2) strategic level which integrates volitional commands and local controllers. Existing technique of parallel computing is very useful. Before we discuss the specifics of how to solve the network related problem on parallel machines, we will first characterize the potentials as well as the limitations of the machines themselves. All computers, serial or parallel, consist of three basic component processors, memory and communication channels. Traditional sequential computers consist of a single central processing unit (CPU), some memory for program and data storage, and I/O channels for communicating between the CPU and memory, and with the outside world. Parallel computers, however, consist of a number of distributed processing units (DPU), memory that can be either shared or distributed among the processors, and channels for communicating among processors and memory as well as with the outside world. As a general rule, the smaller and simpler the individual processors, the more that can be accommodated in a particular machine. This trend provides the basis for classifying parallel machines in two broad categories: fine grained and coarse grained, with fine-grained machines having a very large number of relatively simple processing modes and coarse grained machines having a smaller number of more complex processing modes.

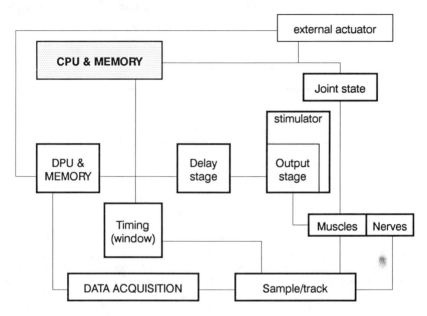

Fig 4. Block diagram of the natural sensor integrated in the artificial reflex control. The signal processing is analog, sample/track is based on sample and hold circuits, preamplifier with input limiters is built in the sample/track part of sensor.

As already said, memory in a parallel machine can be classified as either shared or distributed. Shared memory can be directly accessed by any processor in the machine whereas distributed memory can only by directly accessed by a single processor. Shared memory provides two principal benefits: convenient and efficient means for storing common data that is needed by all of the processors to carry out a computation, and it provides a means for communicating information between processors, without requiring additional dedicated communication channels. The major disadvantage is that contention results if too many processors try to access memory simultaneously. However, distributed memories are easy to access only from a single local controller and redundant communication line has to be designed, but simultaneous memory use is superior in comparison with shared memory.

For our specific purposes we designed a micro-controller using parallel 68HC11 microprocessors. A specific difficulty in this "biological" controllers is induced by so called artifacts. Artifact is, from the point of view of control, noisy feed-back signal. The noise in this case is produced by the control signal it self (Fig 4).

We are using both types of memory, shared and distributed. The 68HC11 microprocessor allows the use of EEPROM in a very effective way, including the possibility to change a content during operation. The connection topology is a so called "hypercube". This specific configuration was selected because the number of connection grows relatively slowly with the number of computers working in parallel

6.CONCLUSION

Sensory driven control plays an important role in the development of active assistive systems. In this paper, a skill based control expert system of this kind based on the study of heuristics of motor skills, has been presented. The proposed approach relies on the transfer of human gait invariants to the controller of a assistive device.

Identification procedure for capturing of invariants of gait characteristics has been developed so that they can be represented as pattern matching rules in the machine.

To assess the validity of the above non-numerical approach in rehabilitation engineering, this control philosophy has been implemented as an active above-knee prosthesis and hybrid orthosis for spinal cord injured paralyzed humans. Clinical tests have proved that skill based expert system control systems reflecting heuristics of motor skills are feasible, efficient and have great inherent potential for improvements.

7. REFERENCES

Berry,F.R.Jr., (1952), Angle Variation Patterns of Normal Hip, Knee and Ankle in Different Operations, University of California, Berkeley, Prosthetic Devices Research Project,Series II,Issue 21

Chomsky,N., (1957), Syntactic structures, Le Haye, Mouton

Kirtley, C. et al., (1985), Influence of Walking Speed on Gait Parameters, J.Biomedical Engineering, Vol 7, 282-288

Nelson, M., Furmaniski,W. and Bower,J., (1989), Simulating neurons and networks on parallel computers, in Methods in Neuronal Modeling, ed Koch and Segev, The MIT Press

Nilsson,J. et al. (1985), Changes in Leg Movements and Muscle Activity with Speed of Locomotion and Mode of Progression in Humans, Acta Physiol. Scand., 123:345-475

Popovic,D. et al., (1986), Technical and Clinical Evaluation of the Self Fitting Modular Orthoses, Final Report, pp 432, NIDRR, Washington, D.C.

Popovic,D. & Schwirtlich,L., (1988), Belgrade Active A/K Prosthesis, in Electrophysiological Kinesiology, J. de Vries, Excerpta Medica, Amsterdam, International Congress Series No 804, 337-343

Popovic,D., Tomovic,R. & Schwirtlich,L., (1989), Hybrid Assistive System - the Motor Neuroprosthesis, IEEE Trans. on Biomedical Engineering, Vol BME-37(7):729-738

Popovic,D., Tomovic,R., Schwirtlich,L. & Tepavac,D., (1990), Control Aspects of an Active A/K Prosthesis, International Journal on Man - Machine Studies, Academic Press, in press

Post,E., (1943), Formal Reduction of the General combinatorial decision problem, Am J. of Mathematics, 65:197-268

Soames,R.W., (1985), Foot Pressure Patterns During Gait, J. Biomedical Engineering, Vol 2:144-149

Tepavac,D., Tomovic,R. & Popovic,D., (1986), Knowledge Base for Reflex Control, Proc. of the XXIX Yugoslav Conference for ETAN, IV.232- 239, Beograd,

Tepavac,D., (1987) Knowledge base for AKP, M.S.Thesis, Faculty of Electrical Engineering, University of Belgrade

Tomovic,R., Popovic,D. & Tepavac,D., (1987), Adaptive Reflex Control of Assistive Systems, in Advances in External Control of Human Extremities, Published by Yugoslav Committee for ETAN, Beograd, 207 - 214,

Tomovic,R., Bekey,G. & Carplus,W., (1987), A Strategy for the Synthesis of Grasp with Multifingered Robot Hands" Proc. of the 1987 Conference IEEE on Robotics and Automation, Vol 3:85-99

Tomovic,R et al. (1988) The Heuristics of Motor Skills, Computer Science Department, UCLA, Technical Report No 880041, Los Angeles, CA

Acknowledgement: This work was partly supported by Research Council of Serbia, Belgrade, Yugoslavia. The author would like to acknowledge Prof Rajko Tomovic for his helpful contribution in general and Dejan Tepavac, M.S. for his work in design of microcomputer programs and knowledge base.

7

PREFERENTIAL
NEURAL NETWORKS

Jozo J. Dujmović

1. INTRODUCTION

The system evaluation problem can be defined as a problem of determining the extent to which a given system fulfills a set of requirements. The final result of an evaluation process should be a scalar indicator of the overall fulfillment of all requirements. This indicator is called the global preference of the evaluated system.

Evaluation, comparison and selection of complex systems is primarily a decision problem, i.e. a logic problem. As indicators of the fulfillment of requirements, preferences are most naturally interpreted as continuous logic variables that may continuously change from a minimum value (0) to a maximum value (1). So, a continuous logic seems to be the most natural environment for solving system evaluation problems.

On the other hand, system evaluation can also be interpreted as a pattern recognition problem. Quantitative criteria for system evaluation are primarily used for detecting such performance patterns that correspond to good systems. Another framework for system evaluation may be the theory of fuzzy sets and systems. If we define a fuzzy set where the greatest grade of membership corresponds to those systems that completely fulfill all requirements defined by a given criterion for system evaluation then the global preference of a system might be interpreted as the grade of membership to such a set.

A natural way to evaluate a complex system (or any compound part of it) consists of (1) decomposing the system into subsystems or components, (2) the separate evaluation of each subsystem, and (3) the aggregation of the subsystem evaluation results. The aggregation process starts with the aggregation of preferences corresponding to the simplest components of the system and terminates when the preferences of the most complex subsystems are aggregated, yielding the global preference of the whole system. A processing unit for preference aggregation is the main component for building complex criterion functions; it can be realized as a specific preferential neuron, and the system evaluation criteria can be organized as neural network models.

This paper presents a class of multi-layered feedforward neural networks called the preferential neural networks (PNN's) and their applications in the area of system evaluation. These networks are based on a continuous valued nonlinear adaptive preferential neuron (ADAPRENE) model. The mathematical basis for preferential neurons is the continuous preference logic (CPL). Each ADAPRENE can be considered a CPL connective and the most important such connectives are described in detail. The training of each ADAPRENE can be done separately and the corresponding criteria and training algorithms are presented. The paper introduces twelve mathematical conditions that the fundamental CPL connective, the generalized conjunction-disjunction (GCD), must satisfy. Five basic models of GCD are presented in the paper, and their specific properties, important for building PNN models, are investigated and compared. Several other conjunctive and disjunctive preferential connectives, known from the literature on fuzzy systems and decision analysis, are reviewed and compared with the GCD model. The problem of the most suitable degree of compensation in compensative connectives is discussed in detail. A special class of ADAPRENE models, based on partial absorption functions, is proposed and their training algorithm is presented. The preferential neural networks for system evaluation are organized as multiple layers of ADAPRENE processing units. Such networks are characterized by a simple stepwise training technique. The input layer of preferential neural networks consists of extended ADAPRENE nodes which include input preference evaluation units, while all other layers consist of ADAPRENE units. The preferential neural networks are successfully applied for evaluation, comparison, selection, and optimization of various computer systems and data management systems. In addition, the ADAPRENE models could be useful as processing units in other neural network models.

Below, we will first introduce the concepts of a continuous preference logic that serves as a mathematical background for organizing the preferential neurons. Then, we will show a technique for realizing the preferential neural networks, as well as the application of such networks in system evaluation.

2.   QUASI-CONJUNCTION, QUASI-DISJUNCTION, AND NEUTRALITY IN THE CONTINUOUS PREFERENCE LOGIC

Each assertion that an evaluated object completely satisfies a given requirement is called a *value statement*. The degree of truth of a value statement is called a *preference*. The continuous preference logic (CPL) can be defined as a continuous logic of value statements.

Evaluated objects can generally have various properties and any selected requirement can be satisfied noway or partially or completely. Accordingly, a value statement can be false, partially true, or completely true. If 0 denotes the truth value of a false statement and 1 denotes the truth value of a true statement then each preference belongs to the interval $I := [0,1]$.

A CPL function is defined as a mapping $I^n \to I$, i.e. the mapping of input preferences $x_1, x_2, \ldots x_n$ into an output preference x_0. In system evaluation models some input preferences are required to affect the output preference more or less than other input preferences, and such a variable degree of importance can be easily adjusted using appropriate weights $W_1, W_2, \ldots, W_n$ ($W_i > 0$, $i = 1, \ldots, n$, $W_1 + W_2 + \ldots + W_n = 1$).

Two fundamental CPL connectives are Quasi-Conjunction (QC) and Quasi-Disjunction (QD) [1]. QC is a connective that implies the coincidence of input preferences, while QD implies the replaceability and compensability of input preferences. In other words, in the case of QC, a relatively high output preference can be obtained only as a consequence of the simultaneous presence of sufficiently high input preferences. A single low input preference in the case of QC can substantially decrease the output preference. Similarly, in the case of QD, the inputs are replaceable, i.e. they are partially equivalent and can substitute each other. A relatively low output preference can be obtained only if all input preferences are simultaneously sufficiently low. Otherwise, a single high input preference in the case of QD can substantially increase the resulting output preference.

The degree of coincidence of high input preferences is called the *conjunction degree*. Using the conjunction degree c the QC connective can be symbolically denoted as follows:

$$x_0 = (W_1 x_1 \; \Delta \; W_2 x_2 \; \Delta \; \ldots \; \Delta \; W_n x_n)^c \; , \quad 0.5 < c < 1 \; .$$

This notation is only a convenient way for systematic specification of all parameters affecting the aggregation of input preferences (i.e. the parameter c is *not* an exponent). The symbol Δ is called **"and"** (and it can be replaced by the word *and*). The conjunction degree c indicates the strength of the quasi-conjunction, i.e. it indicates the similarity between QC and the pure conjunction. QC is similar to the pure conjunction in the sense that the output x_0 is predominantly affected by the lowest values of inputs $x_1, x_2, \ldots, x_n$. Generally, if $0.5 < c < 1$ then the QC satisfies

$$x_1 \wedge x_2 \wedge \ldots \wedge x_n < (W_1 x_1 \; \Delta \; W_2 x_2 \; \Delta \; \ldots \; \Delta \; W_n x_n)^c < (W_1 x_1 + W_2 x_2 + \ldots + W_n x_n)$$

and the boundary conditions are

$$(W_1 x_1 \; \Delta \; W_2 x_2 \; \Delta \; \ldots \; \Delta \; W_n x_n)^{0.5} = W_1 x_1 + W_2 x_2 + \ldots + W_n x_n \; ,$$

$$(W_1 x_1 \; \Delta \; W_2 x_2 \; \Delta \; \ldots \; \Delta \; W_n x_n)^{1} = x_1 \wedge x_2 \wedge \ldots \wedge x_n = \min(x_1, x_2, \ldots, x_n) \; .$$

In a symmetric manner, the degree of replaceability of input preferences is called the *disjunction degree*. Using the disjunction degree d the QD connective can be symbolically denoted similarly to QC:

$$x_0 = (W_1 x_1 \; \nabla \; W_2 x_2 \; \nabla \; \ldots \; \nabla \; W_n x_n)^d \; , \quad 0.5 < d < 1 \; .$$

The symbol ∇ is called **"or"** (and it can be replaced by the word *or*). The disjunction degree d indicates the strength of the quasi-disjunction, i.e. it indicates the similarity between QD and the pure disjunction. QD is similar to the pure disjunction in the sense that the output value x_0 is predominantly affected by the highest values of inputs $x_1, x_2, \ldots, x_n$. If $0.5 < d < 1$ then

$$(W_1 x_1 + W_2 x_2 + \ldots + W_n x_n) < (W_1 x_1 \; \nabla \; W_2 x_2 \; \nabla \; \ldots \; \nabla \; W_n x_n)^d < x_1 \vee x_2 \vee \ldots \vee x_n$$

and the boundary conditions are

$$(W_1 x_1 \; \nabla \; W_2 x_2 \; \nabla \; \ldots \; \nabla \; W_n x_n)^{0.5} = W_1 x_1 + W_2 x_2 + \ldots + W_n x_n \; ,$$

$$(W_1 x_1 \; \nabla \; W_2 x_2 \; \nabla \; \ldots \; \nabla \; W_n x_n)^{1} = x_1 \vee x_2 \vee \ldots \vee x_n = \max(x_1, x_2, \ldots, x_n) \; .$$

The presented properties of QC and QD suggest that the conjunction degree c can be defined so as to represent the normalized average distance (or difference) between QC and the pure disjunction, or the average proximity (or similarity) of the QC and the pure conjunction. Similarly, the disjunction degree d can be defined as the normalized average distance/difference between

QD and the pure conjunction, or the average proximity/similarity of the QD and the pure disjunction. For both QC and QD the arithmetic mean is the boundary function. So, the arithmetic mean separates the QC and QD and accordingly it can be interpreted as the logical neutrality function having neither conjunctive nor disjunctive characteristic properties. The boundary functions for QC and QD are shown in Fig. 1, suggesting that QC is a family of functions between the pure conjunction and the neutrality function, and similarly, QD is a family of functions between the neutrality function (arithmetic mean) and the pure disjunction.

3. THE GENERALIZED CONJUNCTION-DISJUNCTION AND ITS PROPERTIES

The quasi-conjunction and quasi-disjunction can be interpreted as two special cases of a fundamental CPL function called the *generalized conjunction-disjunction* (GCD):

$$x_0 = L(x; q; W) = (W_1 x_1 \diamond W_2 x_2 \diamond \ldots \diamond W_n x_n)^q$$

$$x := (x_1, x_2, \ldots, x_n) , \quad W := (W_1, W_2, \ldots, W_n) , \quad 0 \le q \le 1.$$

The symbol $\diamond$ is called "**and/or**" (and it can be replaced by the word *andor*). The parameter q is called the *degree of compensation*. A low degree of compensation indicates that a low value of any input preference cannot be easily compensated by high values of other input preferences; this property is characteristic for connectives of the QC group. Similarly, a high degree of compensation means that a single high input preference can compensate low values of all other input preferences; that is typical for connectives of the QD group. A symbolic representation of the GCD function is shown in Fig. 2.

The necessary properties of the GCD function can be specified according to various requirements of system evaluation models. A sequence of the twelve most important properties is briefly presented below.

(I) No system can be more desirable than its best component or less desirable than its worst component. Therefore,

$$\min(x_1, x_2, \ldots, x_n) \le (W_1 x_1 \diamond W_2 x_2 \diamond \ldots \diamond W_n x_n)^q \le \max(x_1, x_2, \ldots, x_n) .$$

More precisely, for $0 < q < 0.5$ the GCD function reduces to QC, and for $0.5 < q < 1$ it reduces to QD. For $q = 0$ the GCD function becomes the classical conjunction, and for $q = 1$ it becomes the classical disjunction. Finally, for $q = 0.5$ the GCD function reduces to the neutrality function, i.e. to the weighted arithmetic mean:

$$(W_1 x_1 \diamond W_2 x_2 \diamond \ldots \diamond W_n x_n)^q = x_1 \wedge x_2 \wedge \ldots \wedge x_n , \qquad (q = 0)$$

$$= (W_1 x_1 \triangle W_2 x_2 \triangle \ldots \triangle W_n x_n)^q, \qquad (0 < q < 0.5)$$

$$= W_1 x_1 + W_2 x_2 + \ldots + W_n x_n , \qquad (q = 0.5)$$

$$= (W_1 x_1 \nabla W_2 x_2 \nabla \ldots \nabla W_n x_n)^q \qquad (0.5 < q < 1)$$

$$= x_1 \vee x_2 \vee \ldots \vee x_n \qquad (q = 1)$$

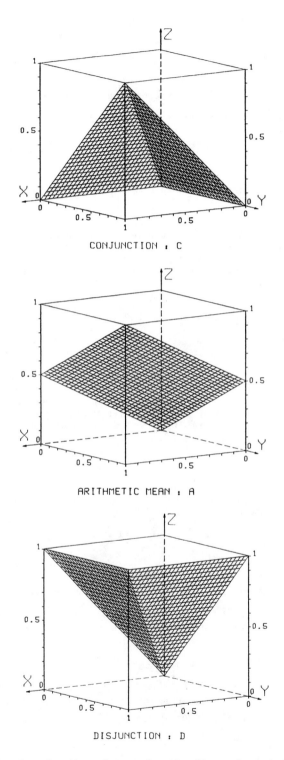

CONJUNCTION : C

ARITHMETIC MEAN : A

DISJUNCTION : D

Figure 1.   Boundary functions for quasi-conjunction and quasi-disjunction

159

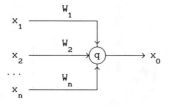

Figure 2. A graphic symbol for generalized conjunction-disjunction

The degree of compensation q can be interpreted as follows:

$$q = d = 1-c \ ,$$

where the range of d and c is now extended to the whole $[0,1]$ interval. In other words, the degree of compensation may be interpreted as the disjunction degree, where the conjunction degree below 0.5 denotes the QD connective and similarly, the disjunction degree below 0.5 denotes the QD connective.

(II)   Idempotency:

$$(\forall z \in [0,1]) \ (W_1 z \lozenge W_2 z \lozenge \ldots \lozenge W_n z)^q = z \ .$$

In this case we also have $\dfrac{\partial L}{\partial q} = 0$, $\dfrac{\partial L}{\partial W_i} = 0$, $i=1,\ldots,n$.

(III)   Any increase of the degree of compensation should yield a corresponding increase of the output preference. In other words, changing the value of q from 0 to 1 it is possible to realize a continuous transition from conjunction to disjunction. If $0 < x_i < 1$, $i=1,\ldots,n$, and all input preferences are *not* equal, then GCD is a strictly increasing function of q:

$$\dfrac{\partial L}{\partial q} > 0 \ .$$

(IV)   The value of a system increases whenever the value of any of its components increases:

$$\dfrac{\partial L}{\partial x_i} > 0 \ , \quad 0<q<1 \ , \quad 0<x_i<1 \ , \quad i=1,\ldots,n \ .$$

(V)   If the preference decrement Δx_i can be compensated by the preference increment Δx_j then the tradeoff between x_i and x_j must be independent of all other input preferences $x_1 ,\ldots,\ x_{i-1}$, $x_{i+1} ,\ldots,\ x_{j-1}$, $x_{j+1} ,\ldots,\ x_n$. In other words, if for given Δx_i we have

$$L(x_1 ,\ldots,\ x_i-\Delta x_i ,\ldots,\ x_j+\Delta x_j ,\ldots,\ x_n; \ q; \ W) = L(x; \ q; \ W) \ ,$$

then Δx_j is not a function of variables $x_1,\ldots,x_{i-1},x_{i+1},\ldots,x_{j-1},x_{j+1},\ldots,x_n.$ This property facilitates the organization of system evaluation models.

(VI)   If an input preference represents a mandatory requirement then its zero value in some cases must cause the zero output preference. Hence, the degree of compensation should have a low compensation threshold value q_0 $(0 \le q_0 < 0.5)$ so that for any compensation degree that is less than or equal to q_0 the following holds:

$$(W_1 x_1 \diamond W_2 x_2 \diamond \ldots \diamond W_n x_n)^q = 0 \ , \ x_i = 0, \ i \in \{1, \ldots, n\}$$
$$> 0 \ , \ x_i > 0, \ i = 1, \ldots, n \ , \ 0 \le q \le q_0 \ .$$

This property holds for q=0, as a consequence of the fundamental property (I). We must note that if $q_0 = 0$ then the pure conjunction is the only function that can be used for modelling various mandatory requirements. That is rather inconvenient since it yields such complex criterion functions that globally can not satisfy the very important property (IV). Therefore, we normally need $q_0 > 0$ and the value $q_0 = 0$ will be referred to as "the trivial low compensation threshold".

A symmetrical high compensation threshold value q_1 $(0.5 < q_1 \le 1)$ can be specified for QD:

$$(W_1 x_1 \diamond W_2 x_2 \diamond \ldots \diamond W_n x_n)^q = 1 \ , \ x_i = 1, \ i \in \{1, \ldots, n\}$$
$$< 1 \ , \ x_i < 1, \ i = 1, \ldots, n \ , \ q_1 \le q \le 1$$

In system evaluation practice complex criteria are predominantly conjunctively polarized and the need for the nontrivial high compensation threshold $(q_1 < 1)$ occurs very rarely. Since the trivial high compensation thresholds $(q_1 = 1)$ are frequently acceptable, the four possible versions of the GCD function can be sorted according to the decreasing desirability as follows:

	Code	The type of GCD function	Low compensation threshold	High compensation threshold
1.	AC	Asymmetrical conjunctive	$q_0 > 0$	$q_1 = 1$
2.	SN	Symmetrical nontrivial	$q_0 > 0$	$q_1 = 1 - q_0$
3.	ST	Symmetrical trivial	$q_0 = 0$	$q_1 = 1$
4.	AD	Asymmetrical disjunctive	$q_0 = 0$	$q_1 < 1$

(VII)   The weight W_i shows the relative importance of the ith input preference. Consequently, the logarithmic input/output sensitivity should be proportional to W_i:

$$\frac{\partial \ln x_0}{\partial \ln x_i} = \frac{x_i}{x_0} \frac{\partial x_0}{\partial x_i} = W_i F_i(x; \ q; \ W) \ , \quad i = 1, \ldots, n \ ,$$

where the value of function F_i should preferably be close to 1 (obviously,

$F_i=1$ only for the geometric mean: $\ln x_0 = W_1 \ln x_1 + \ldots + W_n \ln x_n$).

(VIII) The weights W_i and W_j affect the tradeoff between x_i and x_j ($i \neq j$) as follows:

$$\frac{\partial \ln x_i}{\partial \ln x_j} = -\frac{W_j}{W_i} R_i(x; q; W) \, ,$$

where the function R_i should again be close to 1 (for the geometric mean that is obviously achieved).

(IX) If $W_i=W_j$ then x_i and x_j must be commutative. Otherwise, the pairs $W_i x_i$ and $W_j x_j$ should always be commutative, reflecting the equality of rights of input preferences.

(X) Associativity (see Fig. 3):

$$(W_1 x_1 \diamond W_2 x_2 \diamond W_3 x_3)^q = \left[(W_1 + W_2) \left(\frac{W_1}{W_1 + W_2} x_1 \diamond \frac{W_2}{W_1 + W_2} x_2 \right)^q \diamond W_3 x_3 \right]^q$$

$$= \left[W_1 x_1 \diamond (W_2 + W_3) \left(\frac{W_2}{W_2 + W_3} x_2 \diamond \frac{W_3}{W_2 + W_3} x_3 \right)^q \right]^q$$

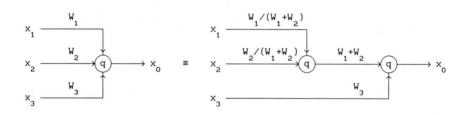

Figure 3. An example of two equivalent preference aggregation structures based on associativity

(XI) Distributivity is defined as follows (see Fig. 4) :

$$[(W_1 x_1 \diamond W_1' (W_2 x_2 \diamond W_2' x_3)^q]^{1-q} = [W_2(W_1 x_1 \diamond W_1' x_2)^{1-q} \diamond W_2'(W_1 x_1 \diamond W_1' x_3)^{1-q}]^q$$

In this definition the primed letters denote negation,

$$W_i' := 1 - W_i \, , \quad i \in \{1, \ldots, n\} \, ,$$

and the negation of the GCD function is denoted as follows:

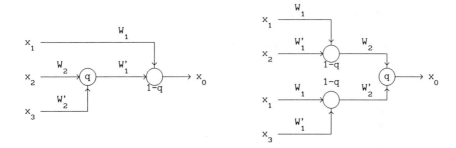

Figure 4. An example of two equivalent preference aggregation structures based on distributivity

$$\left[(W_1 x_1 \Diamond W_2 x_2 \Diamond \ldots \Diamond W_n x_n)^q\right]' = 1 - (W_1 x_1 \Diamond W_2 x_2 \Diamond \ldots \Diamond W_n x_n)^q$$

$$= (W_1 x_1 \Diamond W_2 x_2 \Diamond \ldots \Diamond W_n x_n)^{1-q} .$$

(XII)   De Morgan's Law:

$$(W_1 x_1 \Diamond W_2 x_2 \Diamond \ldots \Diamond W_n x_n)^q = (W_1 x_1' \Diamond W_2 x_2' \Diamond \ldots \Diamond W_n x_n')^{1-q}$$

$$= 1 - [W_1(1-x_1) \Diamond W_2(1-x_2) \Diamond \ldots \Diamond W_n(1-x_n)]^q$$

The presented requirements are empirically shown to be desirable in a variety of system evaluation models. No empirical evidence has been found that additional conditions might be necessary. From the theoretical point of view, however, it would be important to have a set of necessary and sufficient conditions specifying the properties of the GCD function in a rigorous and unique way. The presented sequence of twelve conditions is oriented towards such a goal, but there is no general formal proof that the conditions are either necessary or sufficient. Moreover, some of presented conditions are related (or partially redundant) and it is not difficult to find situations where only a subset of specified conditions may be necessary.

4. THE IMPLEMENTATIONS OF GENERALIZED CONJUNCTION-DISJUNCTION

The desired properties of the GCD function should be used as a guideline for selecting its most appropriate realization. In some applications, where only a subset of properties is necessary, a rather simple implementation of the GCD function can be appropriate. The simplest realization, based on direct combination of conjunction and disjunction, for equal weights and n=2 is the following "minmax version" of GCD:

$$(\frac{1}{2} x_1 \diamond \frac{1}{2} x_2)^q := q \ (x_1 \vee x_2) + (1-q) \ (x_1 \wedge x_2)$$

$$= d \ \max(x_1, \ x_2) + c \ \min(x_1, \ x_2) \ .$$

In the case of three variables,

$$x_0 = (\frac{1}{3} x_1 \diamond \frac{1}{3} x_2 \diamond \frac{1}{3} x_3)^q \ ,$$

assuming that $x_1 \leq x_2 \leq x_3$, the resulting value x_0 can be determined from the feedback network shown in Fig. 5, or from the following algorithm:

WHILE $x_3 - x_1 > \varepsilon$ **DO**
$\quad y_1 := (\frac{1}{2} x_1 \diamond \frac{1}{2} x_2)^q \ ; \ y_2 := (\frac{1}{2} x_1 \diamond \frac{1}{2} x_3)^q \ ; \ y_3 := (\frac{1}{2} x_2 \diamond \frac{1}{2} x_3)^q \ ;$
$\quad x_1 := y_1 \ ; \quad x_2 := y_2 \ ; \quad x_3 := y_3$
END_WHILE ;
$x_0 := x_2$

In this algorithm ε is a selected small value of the resulting error; even for $\varepsilon=0$ this procedure regularly yields x_0 after less than 30 iterations. It can be shown that this algorithm for $x_1 \leq x_2 \leq x_3$ yields the following minmax GDC function of three variables:

$$x_0 = (\frac{1}{3} x_1 \diamond \frac{1}{3} x_2 \diamond \frac{1}{3} x_3)^q = \frac{(1-q)^2 x_1 + q(1-q)x_2 + q^2 x_3}{1 - q + q^2} \ .$$

This approach can be similarly extended for n>3. This version of the GCD function, assuming equal weights, satisfies the conditions I, II, III, IV, V, XI and XII.

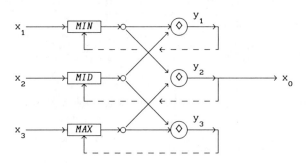

Figure 5. A feedback network for computing the minmax version of GCD for n=3

A substantially better satisfaction of the basic set of 12 conditions can be achieved if the GCD is based on the following weighted power means:

$$M(x; r; W) := \lim_{s \to r} (W_1 x_1^s + W_2 x_2^s + \ldots + W_n x_n^s)^{1/s} , \quad r \in \mathbb{R}_\infty ,$$

$$0 < W_i < 1, \quad i=1,\ldots,n , \quad \sum_{i=1}^{n} W_i = 1 .$$

Weighted power means are the generalization of classic harmonic, geometric, arithmetic and square means (see [GIN58, MIT69, MIT77]); for n=2, and inputs x and y, these classic means in the case of equal weights are

$$h := 2xy/(x+y) ,$$

$$g := \sqrt{xy} ,$$

$$a := (x+y)/2 ,$$

$$s := \sqrt{(x^2 + y^2)/2} , \quad x \geq 0 , \quad y \geq 0 .$$

From these definitions it is easy to note that $g = \sqrt{ah}$ and $a = \sqrt{(g^2 + s^2)/2}$. One of the proofs that $a \geq g$ is $(x + y)/2 - \sqrt{xy} = (\sqrt{x} - \sqrt{y})^2/2 \geq 0$. To prove that $h \leq g$ we first from $g \leq a$ for $a > 0$ find that $g^2/a \leq g$, and then from $g^2 = ah$ we have $h = g^2/a \leq g$. Similarly, from $a^2 \geq g^2 = 2a^2 - s^2$ it follows $a \leq s$. Therefore,

$$h \leq g \leq a \leq s ,$$

and the equality holds only if x=y.

In the general case of n variables and unequal weights the most important special cases of M(x; r; W) are:

$$M(x; r; W) = \min(x_1,\ldots,x_n) , \qquad\qquad r = -\infty$$

$$= 1/(W_1/x_1 + W_2/x_2 + \ldots + W_n/x_n) , \qquad r = -1$$

$$= x_1^{W_1} x_2^{W_2} \ldots x_n^{W_n} , \qquad\qquad r = 0$$

$$= W_1 x_1 + W_2 x_2 + \ldots + W_n x_n , \qquad\qquad r = 1$$

$$= (W_1 x_1^2 + W_2 x_2^2 + \ldots + W_n x_n^2)^{1/2}, \qquad r = 2$$

$$= \max(x_1,\ldots,x_n) , \qquad\qquad r = +\infty$$

and if all input preferences $x_1,\ldots,x_n$ are not equal then M(x; r; W) is an increasing function of the parameter r.

For equal weights $\bar{W} := (1/n, \ldots, 1/n)$ the mean value of $M(x; r, \bar{W})$ is

$$\bar{M}(r) := \int_0^1 dx_1 \int_0^1 dx_2 \ldots \int_0^1 M(x; r; \bar{W}) \, dx_n \quad .$$

Since (see [DUJ73a])

$$\bar{M}(-\infty) := \int_0^1 dx_1 \int_0^1 dx_2 \ldots \int_0^1 \min(x_1, \ldots, x_n) \, dx_n = \frac{1}{n + 1} \quad ,$$

$$\bar{M}(+\infty) := \int_0^1 dx_1 \int_0^1 dx_2 \ldots \int_0^1 \max(x_1, \ldots, x_n) \, dx_n = \frac{n}{n + 1} \quad ,$$

it follows that

$$\frac{1}{n + 1} \leq \bar{M}(r) \leq \frac{n}{n + 1} \quad .$$

So, as proposed in [DUJ74a], the compensation degree q can be defined as the normalized average difference between $\bar{M}(r)$ and $\bar{M}(-\infty)$:

$$q := \frac{\bar{M}(r) - \bar{M}(-\infty)}{\bar{M}(+\infty) - \bar{M}(-\infty)} = \frac{(n+1) \, \bar{M}(r) - 1}{n - 1} =: Q(r,n) \, , \qquad 0 \leq q \leq 1 \quad .$$

From $q = Q(r,n)$, $n > 1$, we can determine $r_n := Q^{-1}(q,n)$, and the corresponding CPL implementation of the AC version of the GCD function is

$$(W_1 x_1 \Diamond W_2 x_2 \Diamond \ldots \Diamond W_n x_n)^q := \lim_{s \to Q^{-1}(q,n)} \left(\sum_{i=1}^n W_i x_i^s \right)^{1/s} \quad . \tag{1}$$

This implementation has almost all the desirable properties: it exactly satisfies the conditions I, II, III, IV, V, VI, VII, VIII, and IX. It also approximately satisfies the conditions X, XI, and XII. An analysis of the corresponding average absolute errors can be found in [DUJ75b]: for associativity (X) the error is less than 1%, for distributivity (XI) a typical error is 1.4%, and for De Morgan laws the average absolute error is regularly less than 4%. For all applications this accuracy is sufficient.

The relationship between the parameters r, c, and d (or q) for all special cases of the GCD function is shown in Fig. 6. Since q is a function of both r and n, and $Q^{-1}(q,n)$ cannot be determined analytically, it is necessary to numerically determine the values of $r_n := Q^{-1}(q,n)$ corresponding to equidistant values of q in the range [0, 1]. Two levels of granularity, where the increments of q are 1/16 and 1/24, are particularly suitable. These

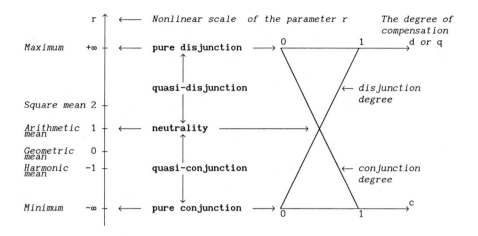

Figure 6. The parameters of the GCD function based on weighted power means

special cases of GCD are given the specific names shown in Table 1. In addition to 17 or 25 variants of GCD corresponding to equidistant values of conjunction and disjunction degrees (c and d) Table 1 also includes as references the classic square, geometric, and harmonic means. Characteristic examples of QC and QD in the case of two variables for the minmax version of GCD are shown in Fig. 7 and for the AC version of GCD based on weighted power means the corresponding examples of QC and QD are shown in Fig. 8 (for equal weights) and in Fig. 9 (for unequal weights). It is easy to note that if $r_n \leq 0$ then any zero input preference yields zero output preference. Therefore, if $q \leq q_0 = Q(0, n)$ then all input preferences represent mandatory requirements. In the case of quasi-disjunction, however, only for q=1 an input preference equal to 1 will yield the value 1 of the output preference. This asymmetric property is desirable in the majority of system evaluation models. For example, if n=2, q=1/3 and q=2/3 then, according to Table 1, $Q^{-1}(1/3, 2)=0$, $Q^{-1}(2/3, 2)=2.504$, and from the definition (1) we have

$$(W_1 x_1 \Diamond W_2 x_2)^{1/3} = x_1^{W_1} x_2^{W_2} \quad ,$$

$$(W_1 x_1 \Diamond W_2 x_2)^{2/3} = \left(W_1 x_1^{2.504} + W_2 x_2^{2.504} \right)^{1/2.504} \quad ,$$

$$(\forall\ x_2 \geq 0)\ (W_1 0 \Diamond W_2 x_2)^{1/3} = 0\ ;\qquad (\forall\ x_1 \geq 0)\ (W_1 x_1 \Diamond W_2 0)^{1/3} = 0\ ,$$

$$(\forall\ x_2 < 1)\ (W_1 1 \Diamond W_2 x_2)^{2/3} < 1\ ;\qquad (\forall\ x_1 < 1)\ (W_1 x_1 \Diamond W_2 1)^{2/3} < 1\ .$$

Table 1
GENERALIZED CONJUNCTION-DISJUNCTION (17 LEVELS)

Name of Operation	Symbol of operation	c, 1-q	d, q	r_2	r_3	r_4	r_5
DISJUNCTION	D	0.0000	1.0000	$+\infty$	$+\infty$	$+\infty$	$+\infty$
STRONG QD (+)	D++	0.0625	0.9375	20.63	24.30	27.11	30.09
STRONG QD	D+	0.1250	0.8750	9.521	11.095	12.27	13.235
STRONG QD (-)	D+-	0.1875	0.8125	5.802	6.675	7.316	7.819
MEDIUM QD	DA	0.2500	0.7500	3.929	4.450	4.825	5.111
WEAK QD (+)	D-+	0.3125	0.6875	2.792	3.101	3.318	3.479
WEAK QD	D-	0.3750	0.6250	2.018	2.187	2.302	2.384
SQUARE MEAN	SQU	0.3768	0.6232	2.000			
WEAK QD (-)	D--	0.4375	0.5625	1.449	1.519	1.565	1.596
ARITHMETIC MEAN	A	0.5000	0.5000	1.000	1.000	1.000	1.000
WEAK QC (-)	C--	0.5625	0.4375	0.619	0.573	0.546	0.526
WEAK QC	C-	0.6250	0.3750	0.261	0.192	0.153	0.129
GEOMETRIC MEAN	GEO	0.6667	0.3333	0.000			
WEAK QC (+)	C-+	0.6875	0.3125	-0.148	-0.208	-0.235	-0.251
MEDIUM QC	CA	0.7500	0.2500	-0.720	-0.732	-0.721	-0.707
HARMONIC MEAN	HAR	0.7726	0.2274	-1.000			
STRONG QC (-)	C+-	0.8125	0.1875	-1.655	-1.550	-1.455	-1.380
STRONG QC	C+	0.8750	0.1250	-3.510	-3.114	-2.823	-2.606
STRONG QC (+)	C++	0.9375	0.0625	-9.060	-7.639	-6.689	-6.013
CONJUNCTION	C	1.0000	0.0000	$-\infty$	$-\infty$	$-\infty$	$-\infty$

GENERALIZED CONJUNCTION-DISJUNCTION (25 LEVELS)

Name of Operation	Symbol of operation	c, 1-q	d, q	r_2	r_3	r_4	r_5
DISJUNCTION	DV3 (D)	0.000	1.000	$+\infty$	$+\infty$	$+\infty$	$+\infty$
VERY STRONG QD, LEVEL 2	DV2	0.042	0.958	31.730	37.49	42.01	48.5
VERY STRONG QD, LEVEL 1	DV1	0.083	0.917	15.082	17.703	19.680	21.7
STRONG QD, LEVEL 3	DS3 (D+)	0.125	0.875	9.521	11.095	12.270	13.24
STRONG QD, LEVEL 2	DS2	0.167	0.833	6.734	7.782	8.557	9.169
STRONG QD, LEVEL 1	DS1	0.208	0.792	5.055	5.787	6.322	6.738
MEDIUM QD, LEVEL 3	DM3 (DA)	0.250	0.750	3.929	4.450	4.825	5.111
MEDIUM QD, LEVEL 2	DM2	0.292	0.708	3.119	3.488	3.750	3.946
MEDIUM QD, LEVEL 1	DM1	0.333	0.667	2.504	2.760	2.938	3.070
WEAK QD, LEVEL 3	DW3 (D-)	0.375	0.625	2.018	2.187	2.302	2.384
WEAK QD, LEVEL 2	DW2	0.417	0.583	1.622	1.722	1.787	1.833
WEAK QD, LEVEL 1	DW1	0.458	0.542	1.289	1.333	1.361	1.381
ARITHMETIC MEAN	A	0.500	0.500	1.000	1.000	1.000	1.000
WEAK QC, LEVEL 1	CW1	0.542	0.458	0.741	0.708	0.688	0.674
WEAK QC, LEVEL 2	CW2	0.583	0.417	0.500	0.443	0.411	0.388
WEAK QC, LEVEL 3	CW3 (C-)	0.625	0.375	0.261	0.192	0.153	0.129
MEDIUM QC, LEVEL 1	CM1 (GEO)	0.667	0.333	0.000	-0.067	-0.101	-0.121
MEDIUM QC, LEVEL 2	CM2	0.708	0.292	-0.314	-0.363	-0.380	-0.388
MEDIUM QC, LEVEL 3	CM3 (CA)	0.750	0.250	-0.720	-0.732	-0.721	-0.707
STRONG QC, LEVEL 1	CS1	0.792	0.208	-1.282	-1.227	-1.168	-1.118
STRONG QC, LEVEL 2	CS2	0.833	0.167	-2.120	-1.947	-1.806	-1.696
STRONG QC, LEVEL 3	CS3 (C+)	0.875	0.125	-3.510	-3.114	-2.823	-2.606
VERY STRONG QC, LEVEL 1	CV1	0.917	0.083	-6.286	-5.393	-4.780	-4.336
VERY STRONG QC, LEVEL 2	CV2	0.958	0.042	-14.613	-12.09	-10.453	-9.43
CONJUNCTION	CV3 (C)	1.000	0.000	$-\infty$	$-\infty$	$-\infty$	$-\infty$

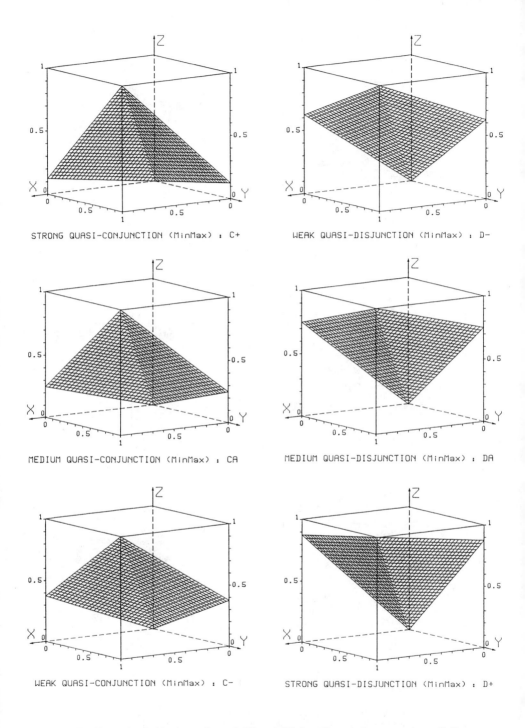

STRONG QUASI-CONJUNCTION (MinMax) : C+

WEAK QUASI-DISJUNCTION (MinMax) : D−

MEDIUM QUASI-CONJUNCTION (MinMax) : CA

MEDIUM QUASI-DISJUNCTION (MinMax) : DA

WEAK QUASI-CONJUNCTION (MinMax) : C−

STRONG QUASI-DISJUNCTION (MinMax) : D+

Figure 7. Characteristic examples of QC and QD for the minimax version of the GCD (two input preferences and equal weights)

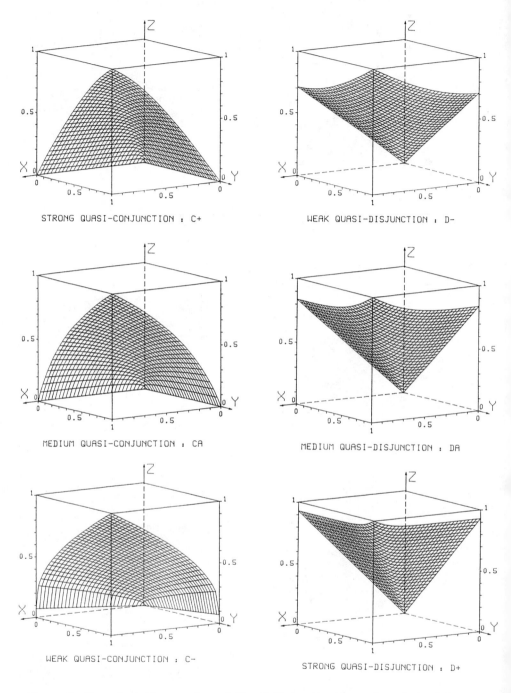

Figure 8. Characteristic examples of QC and QD for the AC version of the GCD function based on weighted power means (n=2, $W_1 = W_2 = 0.5$)

170

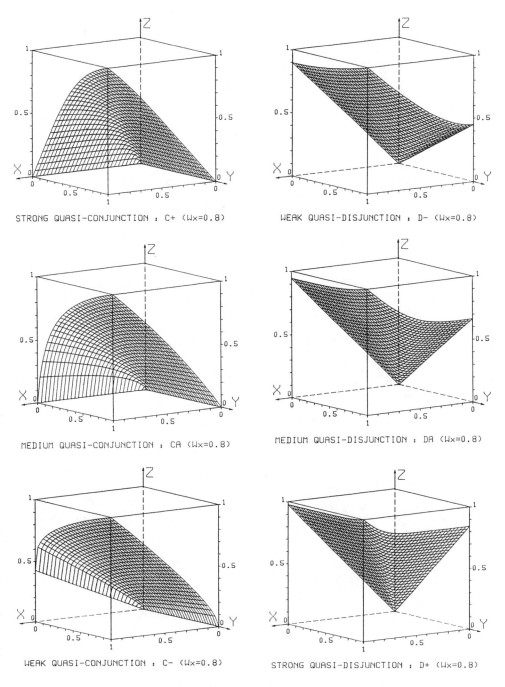

STRONG QUASI-CONJUNCTION : C+ (Wx=0.8)

WEAK QUASI-DISJUNCTION : D- (Wx=0.8)

MEDIUM QUASI-CONJUNCTION : CA (Wx=0.8)

MEDIUM QUASI-DISJUNCTION : DA (Wx=0.8)

WEAK QUASI-CONJUNCTION : C- (Wx=0.8)

STRONG QUASI-DISJUNCTION : D+ (Wx=0.8)

Figure 9. Characteristic examples of QC and QD for the AC version of the GCD function based on weighted power means (n=2, W_x = 0.8, W_y = 0.2)

Other realizations of the GCD function are also possible. It is rather easy to define the GCD function so that De Morgan laws hold. A definition that yields the SN version of the GCD function is

$$(W_1x_1 \lozenge W_2x_2 \lozenge \ldots \lozenge W_nx_n)^q \quad := \lim_{s \to Q^{-1}(q,n)} \left(\sum_{i=1}^{n} W_i x_i^s \right)^{1/s} \;, \qquad 0 \leq q \leq 0.5$$

$$\tag{2}$$

$$= 1 - \lim_{s \to Q^{-1}(1-q,n)} \left[\sum_{i=1}^{n} W_i (1-x_i)^s \right]^{1/s} \;, \quad 0.5 \leq q \leq 1$$

Now all conditions are exactly satisfied except X and XI (the average absolute errors are again low: less than 0.7% for associativity, and less than 1.7% for distributivity). The essential difference between the formulas (1) and (2) is that for (2) the properties of quasi-conjunction and quasi-disjunction are completely symmetrical. If

$$Q^{-1}(1-q,n) \leq 0$$

then a single input preference equal to 1 causes the output that is also equal to 1. For example, if n=2 then for q=1/3 and q=2/3 from definition (2) we have

$$(W_1x_1 \lozenge W_2x_2)^{1/3} \;=\; x_1^{W_1} x_2^{W_2} \;,$$

$$(W_1x_1 \lozenge W_2x_2)^{2/3} \;=\; 1 - (1-x_1)^{W_1} (1-x_2)^{W_2} \;.$$

Obviously, for any value of x_1 and x_2 it follows

$$(W_1 0 \lozenge W_2 x_2)^{1/3} \;=\; (W_1 x_1 \lozenge W_2 0)^{1/3} \;=\; 0 \;,$$

$$(W_1 1 \lozenge W_2 x_2)^{2/3} \;=\; (W_1 x_1 \lozenge W_2 1)^{2/3} \;=\; 1 \;.$$

The suitability of this property should be assessed for any particular system evaluation model. Some characteristic cases of this form of quasi-disjunction are presented in Fig. 10. They are to be compared with the examples of QD for the AC version of GCD shown in Fig. 8.

The ST version of the GCD function based on weighted power means is

$$(W_1x_1 \lozenge W_2x_2 \lozenge \ldots \lozenge W_nx_n)^q \quad := 1 - \lim_{s \to Q^{-1}(1-q,n)} \left[\sum_{i=1}^{n} W_i (1-x_i)^s \right]^{1/s} \;, \qquad 0 \leq q \leq 0.5$$

$$\tag{3}$$

$$= \lim_{s \to Q^{-1}(q,n)} \left(\sum_{i=1}^{n} W_i x_i^s \right)^{1/s} \;, \qquad 0.5 \leq q \leq 1$$

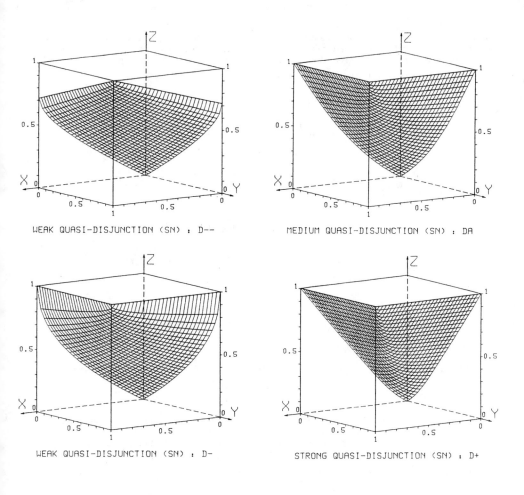

WEAK QUASI-DISJUNCTION (SN) : D--

MEDIUM QUASI-DISJUNCTION (SN) : DA

WEAK QUASI-DISJUNCTION (SN) : D-

STRONG QUASI-DISJUNCTION (SN) : D+

Figure 10. Characteristic examples of QD for the SN version of GCD

The level of errors for associativity and distributivity is again low (less than 1.4%), and other conditions are completely satisfied. The only problem is related to the condition VI where $q_0=0$ and $q_1=1$. For example, from the definition (3) and Table 1 we now have

$$(W_1 x_1 \lozenge W_2 x_2)^{1/3} = 1 - \left[W_1(1 - x_1)^{2.504} + W_2(1 - x_2)^{2.504} \right]^{1/2.504} ,$$

$$(W_1 x_1 \lozenge W_2 x_2)^{2/3} = \left(W_1 x_1^{2.504} + W_2 x_2^{2.504} \right)^{1/2.504} ,$$

$$(\forall\ x_2>0)\ (W_1 0 \lozenge W_2 x_2)^{1/3} > 0 ; \qquad (\forall\ x_1>0)\ (W_1 x_1 \lozenge W_2 0)^{1/3} > 0 ,$$

$$(\forall\ x_2<1)\ (W_1 1 \lozenge W_2 x_2)^{2/3} < 1 ; \qquad (\forall\ x_1<1)\ (W_1 x_1 \lozenge W_2 1)^{2/3} < 1 .$$

The AD version of the GCD function based on weighted power means is

$$(W_1 x_1 \lozenge W_2 x_2 \lozenge \ldots \lozenge W_n x_n)^q := 1 - \lim_{s \to Q^{-1}(1-q,n)} \left[\sum_{i=1}^{n} W_i(1-x_i)^s \right]^{1/s} , \qquad 0 \le q \le 1$$

$$(4)$$

and the asymmetrical properties of this function yield errors in the case of De Morgan laws (XII); the errors of associativity and distributivity are less than 1.5%. In the case of the condition VI the definition (4) yields a disjunctively oriented combination of the compensation thresholds q_0 and q_1; that is easily visible from the examples corresponding to q=1/3 and q=2/3:

$$(W_1 x_1 \lozenge W_2 x_2)^{1/3} = 1 - \left[W_1(1 - x_1)^{2.504} + W_2(1 - x_2)^{2.504} \right]^{1/2.504} ,$$

$$(W_1 x_1 \lozenge W_2 x_2)^{2/3} = 1 - (1 - x_1)^{W_1}(1 - x_2)^{W_2} .$$

$$(\forall\ x_2>0)\ (W_1 0 \lozenge W_2 x_2)^{1/3} > 0 ; \qquad (\forall\ x_1>0)\ (W_1 x_1 \lozenge W_2 0)^{1/3} > 0 ,$$

$$(\forall\ x_2<1)\ (W_1 1 \lozenge W_2 x_2)^{2/3} = 1 ; \qquad (\forall\ x_1<1)\ (W_1 x_1 \lozenge W_2 1)^{2/3} = 1 .$$

This property is regularly not desirable in system evaluation practice since conjunctively polarized criteria are much more frequent than disjunctively polarized criteria.

5. COMPOUND CPL FUNCTIONS

The generalized conjunction-disjunction is the basic CPL function and it can be used as a component for building more complex functions. The compound CPL functions can be organized in a way that is similar to the way the traditional compound Boolean functions are organized by nesting the elementary binary Boolean functions (conjunction, disjunction, and negation). The number of compound CPL functions is greater than in the binary case due to various compensation degrees that can be used in the elementary GCD function. Additional modifications can be realized using different versions of the GCD function.

All fundamental Boolean functions of two variables have their CPL equivalents. For example, the CPL equivalent of the binary exclusive or function

$$x \oplus y \; := \; x \wedge y' \; \vee \; x' \wedge y$$

is the family of functions called the quasi-exclusive disjunction:

$$x \oplus y \; := \; \left[\; W_1(W_2 x \; \Delta \; (1-W_2)(1-y))^c \quad \nabla \quad (1-W_1)(W_3(1-x) \; \Delta \; (1-W_3)y)^c \; \right]^d \;,$$

where the disjunction degree d does not necessarily need to be equal to 1-c. So, the independent parameters of this function are c, d, W_1, W_2, and W_3. They can be selected so as to yield a variety of properties of this function.

The binary equivalence function

$$x \sim y \; := \; x \wedge y \; \vee \; x' \wedge y'$$

can be similarly used to get the following family of quasi-equivalence functions:

$$x \sim y \; := \; \left[\; W_1(W_2 x \; \Delta \; (1-W_2)y)^c \quad \nabla \quad (1-W_1)(W_3(1-x) \; \Delta \; (1-W_3)(1-y))^c \; \right]^d \;.$$

Here we again have five independent adjustable parameters: c, d, W_1, W_2, and W_3.

The binary implication function

$$y \to x \; := \; y' \; \vee \; x$$

yields the following quasi-implication:

$$y \to x \; := \; \left[\; W_1(1-y) \quad \nabla \quad (1-W_1)x \; \right]^d \;,$$

and the adjustable parameters are W_1 and d.

The presented functions of two variables can be expressed using only conjunction and negation:

$$x \oplus y \ := \ ((x \lor y)' \ \lor \ (x' \lor y')')' \ ,$$

$$x \sim y \ := \ (x \lor y)' \ \lor \ (x' \lor y')' \ ,$$

$$y \to x \ := \ y' \lor x \ ,$$

or disjunction and negation:

$$x \oplus y \ := \ (x \land y)' \ \land \ (x' \land y')' \ ,$$

$$x \sim y \ := \ ((x \land y)' \ \land \ (x' \land y')')' \ ,$$

$$y \to x \ := \ (y \land x')' \ .$$

If we replace the conjunction and disjunction in these formulas by the AC version of the GCD function, then, in the case of equal weights, we get six CPL functions depicted in Fig. 11. The asymmetrical low and high compensation thresholds yield different conjunctive and disjunctive versions of the presented quasi-exclusive disjunction, quasi-equivalence, and quasi-implication. The conjunctive versions of these functions are also valid for the binary case $x \in \{0,1\}$, $y \in \{0,1\}$, while the properties of the disjunctive versions somewhat differ.

Some CPL functions are extensions of the classic binary functions, having no direct binary equivalent. For example, in the case of Boolean absorption theorems

$$x \land (x \lor y) = x \ ,$$

$$x \lor (x \land y) = x \ ,$$

the effect of the secondary input y is completely absorbed by the primary input x. Similar theorems do not hold in CPL because the resulting absorption is only partial. However, in CPL we can define the following partial absorption functions

$$\alpha(x, \ y) \ := \ \left(\ W_2 x \ \Delta \ (1-W_2)(W_1 x \ \nabla \ (1-W_1)y)^d \ \right)^c \ ,$$

$$\beta(x, \ y) \ := \ \left(\ W_2 x \ \nabla \ (1-W_2)(W_1 x \ \Delta \ (1-W_1)y)^c \ \right)^d \ ,$$

(5)

with four parameters: W_1, W_2, c, and d. Some versions of these functions are extremely important components of system evaluation models and we are going to present them separately in the following section.

6. THE PARTIAL ABSORPTION FUNCTION

System evaluation models frequently include situations where it is necessary to aggregate preferences of mandatory and desired components of an evaluated system [DUJ80a]. The basic idea of such an aggregation model, called the conjunctive partial absorption (CPA), is exemplified in Fig. 12. The output preference z is a function of the mandatory input x and the desired input y, i.e. the CPA is a mapping $\alpha : \ I^2 \to I$, yielding $z = \alpha(x,y)$. If the mandatory requirement x is not satisfied (i.e. x=0) then the output is z=0 regardless of the value of the desired component y. If x=1 and the desired property y is missing (y=0) then the output preference z is less than x. This case is obtained from the definition (5) if we insert d=0.5 and c=1.

176

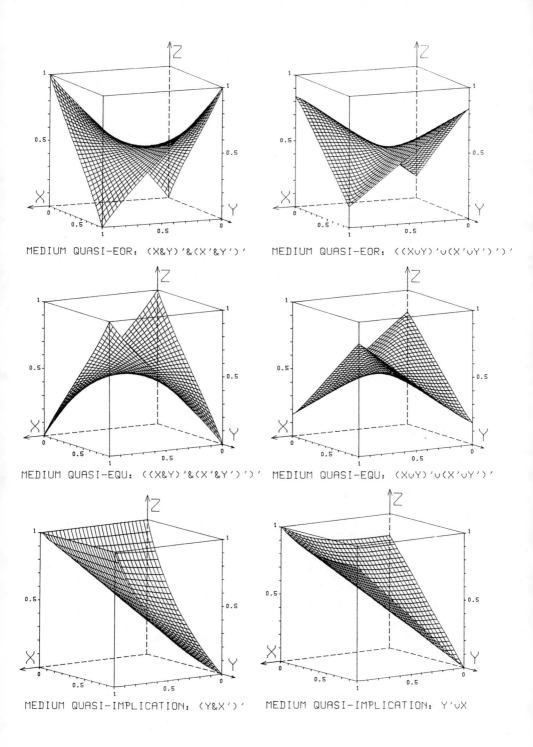

MEDIUM QUASI-EOR: (X&Y)'&(X'&Y')'   MEDIUM QUASI-EOR: ((X∪Y)'∪(X'∪Y')')'

MEDIUM QUASI-EQU: ((X&Y)'&(X'&Y')')'   MEDIUM QUASI-EQU: (X∪Y)'∪(X'∪Y')'

MEDIUM QUASI-IMPLICATION: (Y&X')'   MEDIUM QUASI-IMPLICATION: Y'∪X

Figure 11. Conjunctive and disjunctive versions of the quasi-exclusive
disjunction, quasi-equivalence, and quasi-implication

177

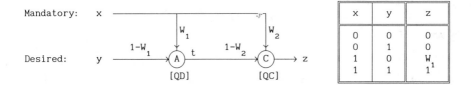

Figure 12. A characteristic "AC" case of conjunctive partial absorption

x	y	z
0	0	0
0	1	0
1	0	W_1
1	1	1

More flexible properties of CPA can be obtained if the pure conjunction is replaced by a quasi-conjunction. The arithmetic mean can be either retained or (with similar results) substituted by a quasi-disjunction. In the case of arithmetic mean and quasi-conjunction the properties of conjunctive partial absorption for various versions of quasi-conjunction are shown in Fig. 13. Such a realization of the conjunctive partial absorption includes three parameters: q, W_1, and W_2. The technique for selecting the most appropriate values of these parameters was proposed in [DUJ80a]. For any value $x \in (0,1)$ if $y>x$ then $z>x$, and similarly, if $y<x$ then $z<x$. The corresponding average increments and decrements, denoted respectively P and M, are defined as follows:

$$P := 2 \int_0^1 \alpha(x,1) \, dx - 1 = 2 \int_0^1 \left(W_2 x \diamond (1-W_2)(W_1 x + 1-W_1) \right)^q dx - 1 ,$$

$$M := [\alpha(x,0)-x]/x = \left[\left(W_2 x \diamond (1-W_2)(W_1 x) \right)^q - x \right] / x = \left(W_2 1 \diamond (1-W_2)W_1 \right)^q - 1.$$

The increment $P(q,W_1,W_2)$ is positive and shows the average improvement of the output preference z with respect to the mandatory input x, as a consequence of the maximum value of the desired input $y=1$. The decrement $M(q,W_1,W_2)$ is negative and shows the average extent to which the output z is less than the mandatory input x, as a consequence of the desired input $y=0$.

The synthesis of the CPA function means the selection of appropriate values of its parameters q, W_1, and W_2. The simplest way to adopt the most suitable values of q, W_1, and W_2 can be organized using the tables of M and P values for various intensities of quasi-conjunction. The values of M and P for $q=0.25$ (i.e. the medium quasi-conjunction CA, $r=-0.72$) are presented in Table 2. Since $-1<M<0$ and $0<P<1$ the values of M and P in Table 2 are multiplied by 100%. For example, if we want the input $y=0$ to cause an average decrement of z by approximately 15%, and $y=1$ to increment z by approximately 5% with respect to x, then, in addition to $q=0.25$, Table 2 suggests the parameters $W_1=0.5$, and $W_2=0.8$. A more accurate approach to the selection of parameters q, W_1, and W_2, using the programming system ANSY for supervised learning, is presented in Section 8.

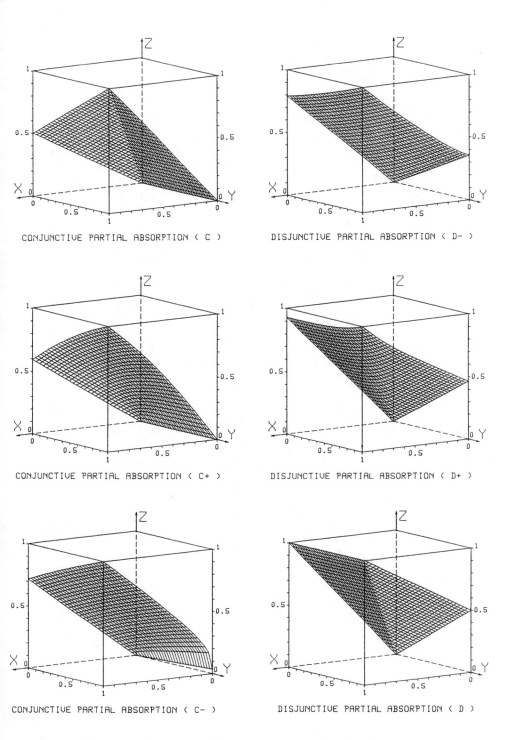

Figure 13. Characteristic cases of the partial absorption function

Table 2

M AND P INDICATORS FOR MEDIUM QUASI-CONJUNCTION

CA		W_1 0.1	0.2	0.3	0.4	0.5	0.6	0.7	0.8	0.9
W_2	0.1	-88.8	-77.9	-67.4	-57.2	-47.1	-37.3	-27.7	-18.3	-9.1
		64.0	57.8	51.5	45.0	38.2	31.3	24.0	16.5	8.5
	0.2	-87.2	-75.5	-64.4	-53.9	-44.0	-35.4	-25.3	-16.6	-8.1
		49.4	45.0	40.4	35.7	30.6	25.4	19.8	13.8	7.3
	0.3	-85.3	-72.5	-60.9	-50.3	-40.5	-31.4	-22.8	-14.8	-7.2
		38.7	35.5	32.1	28.5	24.7	20.6	16.3	11.5	6.2
	0.4	-82.8	-68.8	-56.7	-46.1	-36.6	-28.0	-20.1	-12.9	-6.2
		30.3	27.9	25.4	22.6	19.7	16.6	13.2	9.4	5.1
	0.5	-79.4	-64.2	-51.7	-41.3	-32.3	-24.3	-17.3	-11.0	-5.2
		23.3	21.6	19.7	17.7	15.5	13.1	10.5	7.6	4.2
	0.6	-74.8	-58.2	-45.7	-35.7	-27.4	-20.3	-14.3	-8.9	-4.2
		17.4	16.1	14.8	13.3	11.7	10.0	8.0	5.8	3.3
	0.7	-68.1	-50.4	-38.2	-29.0	-21.8	-16.0	-11.0	-6.8	-3.2
		12.2	11.4	10.5	9.5	8.4	7.1	5.8	4.2	2.4
	0.8	-57.4	-39.6	-28.7	-21.2	-15.6	-11.2	-7.6	-4.6	-2.1
		7.7	7.2	6.6	6.0	5.3	4.6	3.7	2.7	1.6
	0.9	-38.9	-24.0	-16.4	-11.7	-8.3	-5.9	-3.9	-2.4	-1.1
		3.6	3.4	3.1	2.9	2.5	2.2	1.8	1.3	0.8

The CPL function that is symmetrical to the CPA is the disjunctive partial absorption (DPA), exemplified in Fig. 14. It aggregates an input preference x that is sufficient for satisfying the requirement of the DPA criterion, and another input preference, y, that is desired (or necessary) but not sufficient. We define the DPA as a mapping $\beta : I^2 \rightarrow I$, yielding $z = \beta(x,y)$ (cf. the definition (5)). In the case where DPA is based on arithmetic mean (A) and the pure disjunction (D) whenever the sufficient input x is completely satisfied (i.e. x=1) then the output is z=1 regardless of the value of the desired component y. If x=0 and the input preference corresponding to the desired property y is positive then the output preference z is also positive, but less than y. If the pure disjunction is replaced by a quasi-disjunction then the DPA has three adjustable parameters, q, W_1, and W_2, with similar techniques of selecting their values as in the case of CPA; several characteristic cases of this function, for equal weights, are shown in Fig. 13.

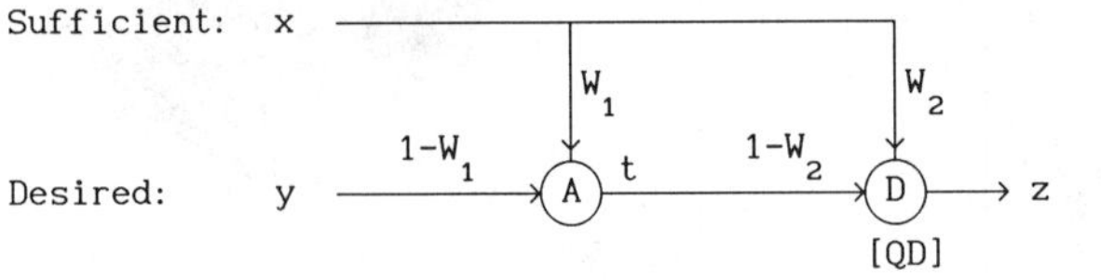

x	y	z
0	0	0
0	1	$1-W_1$
1	0	1
1	1	1

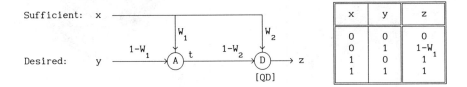

Figure 14. A characteristic "AD" case of the disjunctive partial absorption

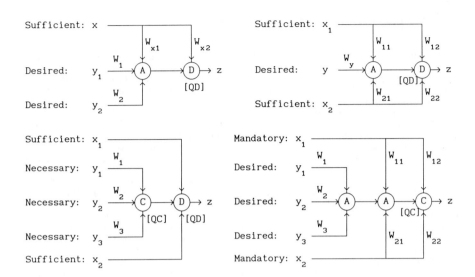

Figure 15. Preference networks with multiple desired, sufficient, mandatory, and necessary conditions

The presented CPA and DPA functions have only two characteristic input values and a single output. The intermediate results, denoted by t, are not used as outputs. Of course, multiple inputs and combined properties are also possible and a number of preference networks with various extensions and/or generalizations can easily be realized. Four such networks are presented in Fig. 15.

The concept of partial absorption can be further extended in the way exemplified in Fig. 16. Now we have three "logical levels" of input preferences denoting mandatory, desired, and optional properties; the corresponding preference network is called the extended conjunctive partial absorption function (ECPA). The presented extension principle can be repeatedly used and the resulting granularity of logical levels can be made arbitrarily fine. It should be noted, however, that in the case of preference networks that are used as decision making models for system evaluation the accuracy of models with either two or three logical levels is regularly sufficient, and finer granularity of presented logical levels is not necessary. Similar extensions are also possible in the case of disjunctive partial absorption.

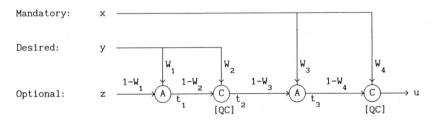

Figure 16. A characteristic case of extended conjunctive partial absorption

181

7. LOGIC CONNECTIVES FOR COMPLEX CRITERION FUNCTIONS

There is a rather long history of research in the area of logic connectives and complex criterion functions. It seems that similar results have emerged in several unrelated research areas: (1) *decision models in operations research and management*, (2) *information retrieval*, (3) *pattern recognition*, (4) *fuzzy systems*, and (5) *neural networks*. An example of applying similar ideas in three of the above areas can be found in [YAG80, YAG87a, YAG87b]. Of course, it is rather difficult to trace the origins of all ideas related to logic connectives for complex criterion functions; this section is aimed only at addressing some relevant issues that may affect the design of preferential neural networks and discussing some approaches that are different from the concepts presented in previous sections.

A detailed study of aggregation connectives in the fuzzy set theory can be found in a paper by Dubois and Prade [DUB85] where the majority of existing connectives is presented in a unifying mathematical way. The terms "aggregation connective" or "logic connective" primarily denote various forms of conjunctive or disjunctive functions as well as complementation (negation). The basic concept of Dubois and Prade is to define, in a form of functional equations, a set of intuitively reasonable axioms that connectives should satisfy, and then to solve these equations in order to derive mathematical expressions characterizing the connectives. That is certainly a correct approach even in cases where the exact satisfaction of all desired properties (or axioms) is not possible or not known. Of course, the meaning of "intuitively reasonable axioms" may substantially differ when defined by those interested in developing a theory, or when defined by those interested in advancing applications.

The fundamental property of logic connectives is the method of moving between the pure conjunction and the pure disjunction. We firmly believe that human mental behavior is *not* based on a nonlinear scale between these extreme functions. For any definition of the distance between the conjunction and the disjunction there must be a parameter showing the current position between them and there must be the arithmetic mean located exactly in the middle. We define the position parameter either as a conjunction degree, or as a disjunction (or compensation) degree. Human decision makers, and educated evaluators in particular, clearly feel the value of that parameter and linearly adjust its value to express the proper type and intensity of the logic connective they want to use. The "linear adjustment" means that the parameter must be defined in such a way that a given change of the parameter always causes the *same* change of properties of the logic connective, independently of the value of the parameter. In other words, the sensitivity of a logic connective to the value of its parameter should be *constant*, and such a scale of parameter values is denoted as "linear". These concepts were initially proposed in [DUJ73b, DUJ74a, DUJ75a] and used to realize a system evaluation method [DUJ73c, DUJ74b] and a theory of complex criteria [DUJ75b]. In all applications we used a linear scale to define the selected position between the conjunction and the disjunction, and the conjunction and disjunction degrees act as parameters that specify the position. Only after the introduction of these parameters we found that evaluators can properly select the desired connective between the *"and"* and the *"or"*. That is not possible if the exponent of power means is directly used as the parameter of logic connectives: the difficulties of proper choice are proportional to the size of the interval of values of the parameter, but a more serious problem is caused by the variable sensitivity of the power mean to the variations of the exponent. Maximum sensitivity is achieved for the arithmetic mean and then the sensitivity nonlinearly decreases as the exponent approaches $+\infty$ or $-\infty$.

The linear scale of the compensation degree is the main difference between our concepts and some concepts proposed by other authors. For example, Dyckhoff and Pedrycz [DYC84] propose the following relationships between the compensation degree q and the weighted power mean parameter r:

$$q(r) = 0.5[1 + r/(1+|r|)] ,$$

or

$$q(r) = 0.5[1 + (2/\pi)arctg(r)] .$$

For r=0 they get q(0)=0.5 suggesting that the geometric mean represents the middle point between the conjunction and the disjunction. That should not be acceptable since it is easy to see that the geometric mean is clearly conjunctively polarized: the simplest version $g=(xy)^{1/2}$ for $x,y\in\{0,1\}$ reduces to the pure conjunction, and from $(xy)^{1/2} < ((x\wedge y) + (x\vee y))/2$, $x\neq y$, $0<x,y<1$, it follows that g is closer to $(x\wedge y)$ than to $(x\vee y)$. Furthermore, the values q(0)=0.5, q(1)=0.75 and q(+∞)=1 suggest that the "compensation difference" between the geometric mean and the arithmetic mean is the same as the compensation difference between the arithmetic mean and the pure disjunction, and that is not acceptable either.

The situation is similar concerning the connective $(x\nabla y)^q(x\Delta y)^{1-q}$ that was used in [ZIM80, ZIM83a, and ZIM83b] in the form

$$x_0 = \left(1 - \prod_{i=1}^{n}(1-x_i) \right)^q \left(\prod_{i=1}^{n} x_i \right)^{1-q} , \qquad 0\leq q\leq 1 .$$

Now, if n=2, $q=x_1=x_2=1/2$ it follows $x_0=\sqrt{3}/4 < 1/2$ showing that for this formula idempotency does not hold, and the central point q=1/2 yields properties that are not neutral, but conjunctively polarized. A similar model,

$$x_0 = q\left(1 - \prod_{i=1}^{n}(1-x_i) \right) + (1-q) \prod_{i=1}^{n} x_i , \qquad 0\leq q\leq 1 ,$$

also proposed in [ZIM80], for n=2, and $x_1=x_2=x$ yields $x_0=x[x+2q(1-x)]$, and idempotency holds only for q=1/2.

Analytic models of complex criterion functions can be found in all areas where it is necessary to use connectives for aggregating quantitative indicators of selected components of a system in order to get a corresponding global indicator for a system as a whole. In [WHI75] J.D. White identifies ten different problem areas where criteria are frequently considered in operations research: transport criteria, educational criteria, marketing criteria, wealth criteria, buildings criteria, production criteria, smoothing criteria, nutritional criteria, public authority criteria, and general economic criteria. Some of these criteria include deterministic situations, and some are used for decision making under uncertainty.

In the case of deterministic management decision models the examples of criterion functions include criteria for selecting the most suitable computer system [JOS68, HIL69, SHA69, GIL71, TIM73, GIL76, DUJ77a, KLE80, DUJ87], or for selecting various consumer goods [MIL70, COU77 SAG77, OST77], for making personal decisions, such as selecting the best career opportunity [MIL66, MIL70], etc. In the case of information retrieval models the criterion functions are organized to derive the extent of matching between the sets of key words and index terms identifying stored records, and some specific request for information expressed by using a specific combination of index terms [SAL83, YAG87b]. In the area of utility theory for management decision making [FIS64, FIS70, KEE76, MAC73, JOH77] the most frequently used models are additive and multiplicative utility functions based on the concept of preference independence [KEE76]. The multiplicative function

$$ku(x_1, \ldots, x_n) + 1 = \prod_{i=1}^{n} [kw_i u_i(x_i) + 1] , \qquad 0 \le u_i \le 1, \ i=1,\ldots,n , \ 0 \le u \le 1$$

for computing the total utility u from the attribute utilities u_i is used in a general case where $w_1 + \ldots + w_n \ne 1$. If $w_1 + \ldots + w_n = 1$ then the utility function has the additive form:

$$u(x_1, \ldots, x_n) = \sum_{i=1}^{n} w_i u_i(x_i) .$$

A well known example of applying such a decision model is the optimum location of an airport, also described in [KEE76]. The basic concept of utility theory is that a tradeoff between preferences is always possible, and such a concept excludes the use of decision models based on conjunction and disjunction. In addition, the presented connectives use a fixed conjunction/disjunction degree.

The above presentation indicates that the most popular logic connectives in applications are based on means. The theory of means [GIN58, MIT69, MIT77] offers a number of mathematical models that might be used for making system evaluation models and their implementation in the form of neural networks. A particularly important family of models can be derived from the weighted quasi-arithmetic means based on a strictly monotone function f:

$$M_f(x; W) := f^{-1}\left(\sum_{i=1}^{n} W_i f(x_i) \right) , \qquad 0 < W_i < 1 , \ i=1,\ldots,n , \qquad \sum_{i=1}^{n} W_i = 1 .$$

Such functions were studied in [DYC85, DOM82a, DOM82b]. A further research should indicate how to select the function f so as to optimize selected properties of logic connectives.

8. ADAPTIVE PREFERENTIAL NEURONS AND THEIR TRAINING

A traditional model of neuron is based on the weighted arithmetic mean and activation and transfer functions shown in Fig. 17 [SOU88]. Such a neuron and a parameter adjustment technique based on the back propagation algorithm for supervised learning can be used to get feedforward neural network systems that are general in the sense of Kolmogorov [HEC87] and that can be applied in a variety of situations [CSS89]. Such networks can also be applied for system evaluation. Unfortunately, in real situations the training of such a network is not possible since we are not able to provide data for the necessary training set. For example, the number of inputs in the case of evaluating a minicomputer or a midicomputer typically varies from 80 to 150. For detailed analyses and/or for more complex systems up to 400 inputs can be identified.

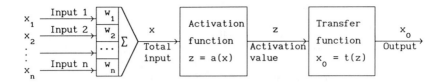

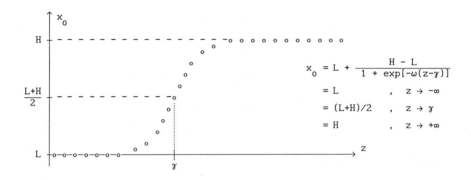

$$x_0 = L + \frac{H - L}{1 + \exp[-\omega(z-\gamma)]}$$

$$= L \qquad , \quad z \to -\infty$$

$$= (L+H)/2 \qquad , \quad z \to \gamma$$

$$= H \qquad , \quad z \to +\infty$$

Figure 17. A traditional model of neuron with a sigmoid transfer function

In the case of system evaluation, neural network models are not used for modelling input-output relationships of man-made systems or similar systems where an arbitrary number of input-output relations can easily be collected and used as a training set. By contrast, all neural networks for system evaluation are used for modelling human evaluation process and the only way to get the training set is directly from human decision makers. Unfortunately, in a typical case with 100 inputs, that means that each component of the training set consists of 101 numbers. Human decision makers, regardless of the level of expertise, neither can directly define such a function of 100 variables with a reasonable confidence level, nor can provide such a number of these functions that would be sufficient to get a meaningful training set. So, we are faced with the problem of preparing a training set whose size and complexity are beyond any acceptable level.

A method to cope with the size and complexity of real system evaluation problems consists of using preferential neurons and training the corresponding PNN separately neuron by neuron. That is possible because the system evaluation networks have a tree-like structure where each neuron works as a preference aggregation unit having well defined inputs and output. In addition, the number of inputs per neuron in practice regularly varies from 2 to 5 and for such a neuron a supervised learning scheme can easily be organized.

The adaptive preferential neuron (ADAPRENE) can be realized using suitable CPL functions and a supervised training technique. The simplest possible ADAPRENE is the GCD-ADAPRENE presented in Fig. 18.

185

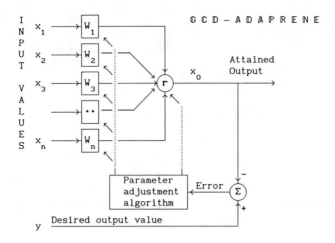

Figure 18. The simplest adaptive preferential neuron: GCD-ADAPRENE

The training of the GCD-ADAPRENE is based on the following training set and error function:

$TRAINING$	SET
INPUT VALUES	OUTPUT
x_{11} x_{12} $\cdots$ x_{1n}	y_1
x_{21} x_{22} $\cdots$ x_{2n}	y_2
. . . .	. . .
x_{k1} x_{k2} $\cdots$ x_{kn}	y_k

Mean square error:

$$\delta(W_1, \ldots, W_n, r) :=$$

$$= \frac{1}{k} \sum_{j=1}^{k} \left[y_j - \left(\sum_{i=1}^{n} W_i x_{ji}^r \right)^{1/r} \right]^2$$

It can easily be noted that the above δ function cannot be directly used as a criterion function and minimized. We must satisfy the additional conditions:

$$W_1 + W_2 + \ldots + W_n = 1 \quad,$$
$$0 < W_i < 1 \quad, \quad i=1,\ldots,n \ .$$

So, if we define an auxiliary criterion function for the sum of weights:

$$\lambda(W_1, W_2, \ldots, W_n) := \mid W_1 + W_2 + \ldots + W_n - 1 \mid \quad,$$

and an auxiliary function for the range of weights:

186

$$\mu(W_i) := 1 \ , \qquad W_i \leq 0$$
$$= 0 \ , \qquad 0 < W_i < 1$$
$$= 1 \ , \qquad W_i \geq 1 \ , \qquad i \in \{1, \ldots, n\} \ ,$$

then the compound criterion function for supervised learning can be defined as follows:

$$\Psi(W_1, W_2, \ldots, W_n, r) := \delta(W_1, W_2, \ldots, W_n, r)$$
$$+ \ \rho \ \lambda(W_1, W_2, \ldots, W_n)$$
$$+ \ \sum_{i=1}^{n} \mu(W_i) \ ,$$

where ρ denotes a parameter used to adjust the influence of the λ function. An efficient way to find the parameters W_1, W_2, ..., W_n, and r is to minimize the Ψ function using the simplex algorithm [NEL64]. The optimum values of the parameters correspond to the minimum value of the Ψ function. The computation of the optimum values of parameters can be done using the programming system ANSY for the analysis and synthesis of CPL functions [DUJ86]. ANSY performs the minimization of the Ψ function using the sequence of parameters

$$\rho = 0.1 \ , \ 1 \ , \ 2 \ , \ 4 \ , \ 8 \ , \ 16 \ .$$

Initially, $\rho = 0.1$. Using that value, ANSY finds the optimum values of parameters W_1^*, W_2^*, ..., W_n^*, r^* that yield the minimum of Ψ. If the training set is defined so that the minimum value of Ψ is zero then in the large majority of cases the obtained results of the minimization (for $\rho = 0.1$) satisfy the condition

$$\Psi(W_1^*, W_2^*, \ldots, W_n^*, r^*) < 10^{-4}$$

and the resulting values W_1^*, W_2^*, ..., W_n^*, and r^* are adopted as parameters of the corresponding ADAPRENE. If $\Psi \geq 0.0001$ then the obtained values W_1^*, W_2^*, ..., W_n^*, and r^* (that are already very close to the resulting minimum) are taken as initial conditions for the next minimization, but now ANSY uses $\rho = 1$ to increase the influence of the λ function. At the end of minimization ANSY tests the condition $W_1^* + W_2^* + \ldots + W_n^* = 1$, and if this condition is not satisfied then the minimization repeats using $\rho = 2$. The training set may contain values that cannot yield zero value of the criterion function Ψ. In such cases ANSY finds parameters that correspond to the minimum of Ψ and the zero value of λ.

Sometimes it is not necessary to determine the values of all parameters: the user may wish to fix and adopt the values of some parameters in advance. Such values are usually known from previous analyses or from experience. For example, if the user knows that the given neuron must use the neutrality function (r=1) then it is necessary to determine only the weights W_1, ..., W_n. ANSY supports such analyses using a parameter vector P whose components are

187

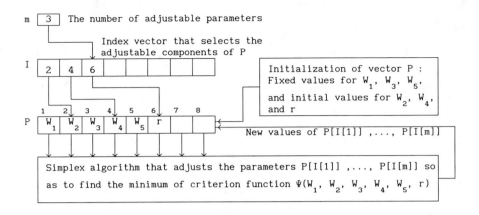

Figure 19. Optimization of adjustable parameters of the criterion function Ψ

$$P[1]=W_1 \ , \quad P[2]=W_2 \ , \quad \ldots \ , \quad P[n]=W_n \quad P[n+1]=r \ ,$$

and an index vector, $I[1..m]$, that is used for identifying the selected components of the parameter vector P. The corresponding optimization technique is exemplified in Fig. 19.

Initial values of n weights can always be taken equal, $W_i=1/n$, $i=1,\ldots,n$. For the exponent r it is useful to detect whether we have the quasi-conjunction (where the output values are lower than the arithmetic mean of input values) or the quasi-disjunction (where the output values from the training set are greater than the arithmetic mean); then we adopt a negative or a positive initial value:

$$r := -r_{init} \ , \qquad \sum_{j=1}^{k} \left(y_j - \frac{1}{n} \sum_{i=1}^{n} x_{ji} \right) < 0$$

$$= +r_{init} \ , \qquad \sum_{j=1}^{k} \left(y_j - \frac{1}{n} \sum_{i=1}^{n} x_{ji} \right) \geq 0$$

Experiments with the simplex algorithm suggest that the initial values should be taken from the interval $2 < r_{init} < 10$ (e.g., ANSY uses $r_{init}=7.7$).

An example of training a GCD-ADAPRENE using ANSY is presented in Fig. 20. Let us also note that if only a single parameter is to be determined then the simplex algorithm cannot be applied; in such cases ANSY uses a separate procedure of one-dimensional minimization.

```
THE SYNTHESIS OF GENERALIZED CONJUNCTION-DISJUNCTION

                  W1
    x1 ------>---------.
                  W2   :
    x2 ------>-------.  :
     .                : :
     .               .---.
     . ------>----->: r :------>  y
     .               '---'
     .              Wn   :
    xn ------>--------'

Enter the number of input variables:   n =  4

This function has the following parameters:

Identifier:    1    2    3    4    5
---------------------------------------
Parameter :   W1   W2   W3   W4    r

How many parameters do you want to determine ?    5

Enter a table containing any number of desired I/O values
(a value outside the [0,1] interval denotes  the end of input data)

x1  x2  x3  x4     y
---------------------
0.5 1.0 1.0 1.0   0.80
1.0 0.5 1.0 1.0   0.85
1.0 1.0 0.5 1.0   0.70
1.0 1.0 1.0 0.5   0.90
999 999 999 999   9999


THE RESULTING OPTIMUM VALUES OF PARAMETERS

W1 =   0.259443
W2 =   0.183951
W3 =   0.440307
W4 =   0.116299
r  =  -0.861742

The minimum value of criterion function = 0.575E-07
The number of function computations =   337
W1+...+Wn = 0.1000E+01,  Max error of result = 0.936E-06

===================================================
Given values                    Analytic model
---------------------------------------------------
 x1    x2    x3    x4     y        y      Error
---------------------------------------------------
0.50 1.00 1.00 1.00   0.8000   0.8000  -0.19E-07
1.00 0.50 1.00 1.00   0.8500   0.8500   0.14E-07
1.00 1.00 0.50 1.00   0.7000   0.7000  -0.65E-08
1.00 1.00 1.00 0.50   0.9000   0.9000  -0.93E-07
===================================================
```

Figure 20.   A sample ANSY output showing the training of a GCD-ADAPRENE

189

The presented GCD-ADAPRENE is not the only ADAPRENE that can be used in PNN's. All other CPL functions can also be used as PNN components. The most important of these functions in system evaluation models is the general four-parameter partial absorption function:

$$z = \pi(x,\ y\ ;\ W_1,\ W_2,\ s,\ r) := \left[\ W_2 x^r + (1-W_2)\left(\ W_1 x^s + (1-W_1)y^s\ \right)^{r/s}\ \right]^{1/r} ,$$

$$0 < W_1 < 1\ ,\quad 0 < W_2 < 1\ ,\quad s \in \mathbb{R}_\infty\ ,\quad r \in \mathbb{R}_\infty\ .$$

Using this function the PA-ADAPRENE can be organized as shown in Fig. 21.

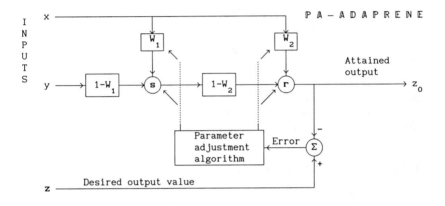

Figure 21. PA-ADAPRENE: a preferential neuron based on partial absorption

Due to the independence of the PA-ADAPRENE parameters the training is now simpler than in the case of the GCD-ADAPRENE (there is no need for the λ function). Therefore, the training set and the compound criterion function Ψ can be organized as follows:

INPUT	VALUES	OUTPUT
x_1	y_1	z_1
x_2	y_2	z_2
. . .	. . .	. . .
x_k	y_k	z_k

T R A I N I N G   S E T

$$\Psi(\ W_1,\ W_2,\ s,\ r\ ) :=$$

$$= \frac{1}{k} \sum_{j=1}^{k} \left[\ z_j - \pi(x_j,\ y_j\ ;\ W_1,\ W_2,\ s,\ r)\ \right]^2$$

$$+ \mu(W_1) + \mu(W_2)\ .$$

```
THE SYNTHESIS OF PARTIAL ABSORPTION FUNCTION
```

```
This function has the following parameters:
Identifier:   1   2   3   4
------------------------------
Parameter :  W1  W2   s    r

How many parameters do you want to determine? 3
Enter the identifiers of selected parameters: 1  2  4

Enter fixed values of remaining parameters:
s  = 1

Enter a table containing any number of desired x,y,z values
(a value outside the [0,1] interval denotes the end of data)

x    y    z
-----------
0    1    0
1    0    .8
.5   1    .55
99   99   999

THE RESULTING OPTIMUM VALUES OF PARAMETERS

W1 =   0.375870
W2 =   0.793529
r  =  -0.256535

The minimum value of criterion function = 0.371E-08
The number of function computations  = 150
The achieved maximum error of result = 0.000001

=======================================================
Triplets of desired values  :    Analytical  model
----------------------------+--------------------------
    x      y      z         :     z        Error
----------------------------+--------------------------
  0.000  1.000  0.0000      :   0.0000   0.0000000
  1.000  0.000  0.8000      :   0.8000   0.0000000
  0.500  1.000  0.5500      :   0.5500   0.0000000
=======================================================

Do you have other desired values for this function (Yes/No) ?  no
Have you completed the synthesis of the PA function (Yes/No) ? yes
```

Figure 22.   A sample ANSY output showing the training of a PA-ADAPRENE

The minimization of $\Psi(W_1, W_2, s, r)$ can be performed using the technique presented in Fig. 19 where

$$P[1]=W_1 \ , \quad P[2]=W_2 \ , \quad P[3]=s \ , \quad P[4]=r$$

and the simplex algorithm is applied for minimizing the criterion function Ψ where the parameters $P[I[1]],\ldots,PI[[m]]$, $(1 \le m \le 4)$, are adjustable, and the remaining parameters are fixed. An example of the PA-ADAPRENE training using the programming system ANSY is presented in Fig. 22.

Other ADAPRENE units can be realized similarly to the GCD-ADAPRENE and the PA-ADAPRENE. In the case of system evaluation models, however, the GCD-ADAPRENE and the PA-ADAPRENE are the most frequently used processing units. More precisely, the preference aggregation functions used by the ADAPRENE units in system evaluation models can be sorted according to the decreasing frequency of use as follows:

(1) *quasi-conjunction,*
(2) *neutrality,*
(3) *conjunctive partial absorption,*
(4) *quasi-disjunction,*
(5) *extended conjunctive partial absorption,*
(6) *disjunctive partial absorption,*
(7) *conjunction,*
(8) *disjunction.*

As an example of other possible ADAPRENE units, Fig. 23 shows a unit for quasi-equivalence and quasi-exclusive disjunction that uses five independently adjustable parameters: W_1, W_2, W_3, r, and s. In addition to GCD functions this unit includes two inverters.

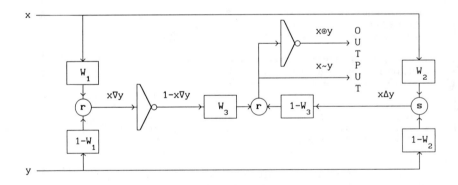

Figure 23.   An ADAPRENE unit for quasi-equivalence and quasi-exclusive
             disjunction

9. PREFERENTIAL NEURAL NETWORKS FOR SYSTEM EVALUATION, COMPARISON, AND OPTIMIZATION

Preferential neural networks for system evaluation regularly have the global tree structure where nodes represent specific ADAPRENE units performing preference aggregation operations. Suppose, for example, a situation where several computer systems have to be evaluated and compared. The global preference of a computer system depends on subsystem preferences corresponding to hardware, software, measured performance, and vendor support, as shown in Fig. 24. In other words, the requirements that should be satisfied by a computer system as a whole can be decomposed and expressed as separate requirements for hardware, software, performance, and support. Similarly, the requirements corresponding to hardware can be further decomposed into subsystem requirements corresponding to the central processing unit, external memory, I/O units, and network facilities. Such a decomposition process obviously yields a system requirement tree structure and terminates when sufficiently simple requirements for individual components or attributes of the evaluated system are generated. These final components for evaluation are called *performance variables*. The fundamental property of performance variables is that for each performance variable we can define a function that directly maps a given value of performance variable into the corresponding value of preference, called the *elementary preference*. The mapping reflects the requirements specified by the evaluator and represents a part of the system evaluation model. The set of performance variables is assumed to include *all* attributes that affect the global preference of each evaluated system.

The preference aggregation process is characterized by the preferential independence property: the tradeoffs between subsystem preferences at a given aggregation node do not depend on the level of remaining preferences. For example, a tradeoff policy between computer hardware and measured performance does not depend on the quality of vendor support. Assuming that form of preferential independence it is possible to separately define each preference aggregation function. If the global criterion function is realized as a preferential neural network then the adjustment of parameters of individual neurons can be done separately for each neuron; such a process will be called the *stepwise supervised training.*

Let the vector of performance variables be $(x_1, \ldots, x_n)$ and let $x_i \in X_i \subset \mathbb{R}$, $i=1,\ldots,n$. Here X_i represents the range of possible values of x_i and the corresponding *performance variable space*, $X := \prod_{i=1}^{n} X_i$, represents a rectangular subset of a finite-dimensional Euclidean space. So, our task is to define a global preference evaluation function over X, i.e. our preferential neural network should realize the mapping $G: X \to [0,1]$ that yields the *global preference* of the evaluated system $E_0 := G(x_1, x_2, \ldots, x_n)$.

The values of performance variables $x_1, x_2, \ldots, x_n$ are real numbers. Therefore, before using preferential neurons we have to map the values of performance variables into elementary preferences. The corresponding *elementary criterion functions* $g_i: X_i \to [0,1]$ then yield elementary preferences $E_i := g_i(x_i)$, $i=1,\ldots, n$. An early example of elementary criterion functions can be found in [WHI63], an excellent survey of various criterion functions was

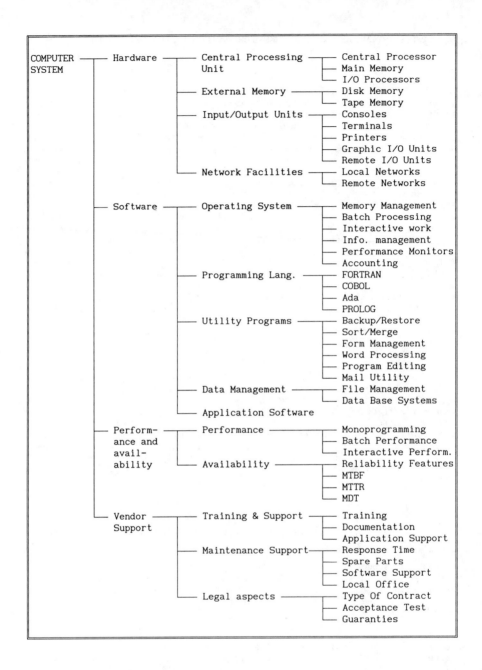

Figure 24. The first three levels of a system requirement tree for computer evaluation

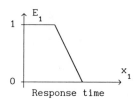 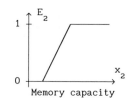 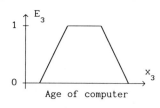

| Response time | Memory capacity | Age of computer |

Figure 25. Characteristic examples of elementary criterion functions

prepared by J. R. Miller in [MIL66] and [MIL70], and a systematic classification of 12 different types of elementary criteria was proposed in [DUJ82].

Three characteristic examples of elementary criterion functions are shown in Fig. 25. In the first example we evaluate the measured response time of a terminal. If the response time is less than a given low threshold value we are perfectly satisfied, i.e. E_1=1. If the response time exceeds a high threshold value then we cannot accept such a system and E_1=0. Between the two thresholds we use a linear interpolation. A similar approach is applied in the second example where we evaluate a proposed memory capacity (x_2), but now we are perfectly satisfied if the proposed memory is greater than or equal to a high threshold value. The third example shows a selective function for the evaluation of the age of a computer. Such a function is aimed at discouraging salesmen from proposing either premature and insufficiently verified computers, or too old computers that may soon be out of production.

Using the elementary criterion functions we get the elementary preferences E_1, E_2,..., E_n that are then used as inputs in the preference aggregation process. Therefore, the preferential neurons in the first layer of a preferential neural network have the structure exemplified in Fig. 26. The output of the first layer consists of preferences that are then used as inputs for the second layer of PNN. The number of output preferences decreases from layer to layer yielding at the last level a single global output preference:

FIRST LAYER	Performance variables:	x_1 x_2 $\cdots$ x_n	
	Elementary preferences:	E_1 E_2 $\cdots$ E_n	
	First layer output preferences:	E_{11} E_{12} $\cdots$ E_{1n_1}	$(n_1 < n)$
SECOND LAYER	Second layer output preferences:	E_{21} E_{22} $\cdots$ E_{2n_2}	$(n_2 < n_1)$
	$\cdots \cdots \cdots \cdots \cdots \cdots$		
LAST LAYER	Global output preference:	E_0	

The preferential neurons in the second layer, as well as in all subsequent layers, differ from the first layer neurons because they only consist of appropriate CPL connectives. The elementary criteria are included only in the first layer.

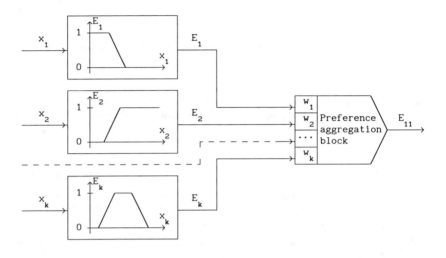

Figure 26. A typical preferential neuron in the input layer

The presented procedure for the synthesis of PNN's for system evaluation can now be summarized as the following three main steps:

STEP 1:   *Development of the system requirement tree*

Starting from the initial node labeled "system as a whole" we systematically decompose requirements and generate lower level nodes representing subcategories of requirements. This process terminates when elementary requirements for individual performance variables are generated.

STEP 2:   *Definition of elementary criteria*

For each performance variable x_i, an elementary criterion function g_i is defined and used for computing the corresponding elementary preference $E_i = g_i(x_i)$, $0 \leq E_i \leq 1$, $i=1,\ldots,n$.

196

STEP 3: *Aggregation of preferences*

The elementary preferences E_1, E_2, ..., E_n are aggregated using a multy-layer network of individually trained preferential neurons. The preference aggregation network realizes the preference aggregation function $A: I^n \to I$ and yields the global preference of the evaluated system:

$$
\begin{aligned}
E_0 &= A(E_1, E_2, \ldots, E_n) \\
&= A(g_1(x_1), g_2(x_2), \ldots, g_n(x_n)) \\
&= G(x_1, x_2, \ldots, x_n) \quad .
\end{aligned}
$$

Let us now present a short example illustrating the above procedure. The example will show a PNN for job selection (a similar example was used by J. R. Miller to illustrate an additive scoring model proposed in [MIL66] and [MIL70]). The system requirement tree yielding twelve performance variables can be specified as follows:

OFFERED JOB

1. Monetary compensation   *Performance*
    1.1 Starting salary and fringe benefits   *variables*
        1.1.1   Starting salary ————————————→   (1)
        1.1.2   Fringe benefits ————————————→   (2)
    1.2 Anticipated future salary   ————————→   (3)

2. Main characteristics of the job
    2.1 Immediate training requirements   ————————→   (4)
    2.2 Attractiveness of job
        2.2.1   Technical content of the job   ————→   (5)
        2.2.2   Obtained training and experience ————→   (6)

3. Total travel requirements
    3.1 Daily commuting requirements   ————————→   (7)
    3.2 Business traveling
        3.2.1   Maximum trip length   ————————→   (8)
        3.2.2   Annual percentage of time away from home ——→   (9)

4. Location of job
    4.1 Climate and environment   ————————————→   (10)
    4.2 Degree of urbanity   ————————————————→   (11)
    4.3 Proximity to relatives   ————————————→   (12)

The obtained performance variables can be evaluated using the elementary criteria presented in Fig. 27. Each elementary criterion includes a preference scale and its description. Preferences are presented in the range from 0 to 100%. The preference scales show the values of performance variables that correspond to some selected values of preference. For example, if the daily commuting requirements (elementary criterion #300) are less than or equal to 30 minutes we are perfectly satisfied and the assigned preference value is 100%. If the total time spent daily in commuting is 3 hours or more, then the assigned preference value will be 0. A linear interpolation is used for computing all other values of preference (e.g. the commuting time of 1.75 hours yields the preference value of 50%). Some elementary criteria use

L S P — **E L E M E N T A R Y   C R I T E R I A**

100 — STARTING SALARY
Evaluated as the following relative salary:
$$SREL := 100*S/SMAX,$$
where:
S := offered annual salary decreased by taxes and locally adjusted to account for different living costs at various geographical locations,
SMAX := the maximum offered S.

Scale: 100–100, 90–, 80–, 70–, 60–, 50–, 40–, 30–, 20–, 10–, 0–50

110 — FRINGE BENEFITS
Evaluated as the following relative fringe benefits:
$$FREL := 100*F/FMAX,$$
where:
F := locally adjusted after-tax annual insurance benefits plus locally adjusted after-tax annual retirement benefits,
FMAX := the maximum offered F.

Scale: 100–100, 90–, 80–, 70–, 60–, 50–, 40–, 30–, 20–, 10–, 0–50

120 — ANTICIPATED FUTURE SALARY
Evaluated as the following relative anticipated salary:
$$AREL := 100*A/AMAX$$
where:
A := 0.5 * (anticipated 3-year salary) + 0.5*(anticipated 5-year salary)
AMAX := the maximum offered A. (Each anticipated salary is defined as the locally adjusted after-tax annual salary.)

Scale: 100–100, 90–, 80–, 70–, 60–, 50–, 40–, 30–, 20–, 10–, 0–50

200 — IMMEDIATE TRAINING REQUIREMENTS
Necessary initial training as a preparation for the proposed job. (Expressed in months)

Scale: 100–0, 90–, 80–, 70–, 60–, 50–, 40–, 30–, 20–, 10–, 0–12

210 — TECHNICAL CONTENT OF THE JOB
Direct preference assessment of the personal interest in the technical content of the job and the degree of variety implicit in the job.

Scale: 100–100, 90–, 80–, 70–, 60–, 50–, 40–, 30–, 20–, 10–, 0–0

220 — OBTAINED TRAINING AND EXPERIENCE
Direct preference assessment of the amount of training and experience realizable from the offered job.

Scale: 100–100, 90–, 80–, 70–, 60–, 50–, 40–, 30–, 20–, 10–, 0–0

300 — DAILY COMMUTING REQUIREMENTS
Total time spent daily in commuting, expressed in hours.

Scale: 100–0.5, 90–, 80–, 70–, 60–, 50–, 40–, 30–, 20–, 10–, 0–3

310 — MAXIMUM TRIP LENGTH
Maximum duration of an extended business trip expressed in days.

Scale: 100–0, 90–, 80–, 70–, 60–, 50–, 40–, 30–, 20–, 10–, 0–30

320 — ANNUAL PERCENTAGE OF TIME AWAY FROM HOME
The percentage of total time spent yearly on business trips.

Scale: 100–0, 90–, 80–, 70–, 60–, 50–, 40–, 30–, 20–, 10–, 0–50

400 — CLIMATE AND ENVIRONMENT
Direct preference assessment of the climate in the region of the offered job and an assessment of the total attractiveness of the region.

Scale: 100–100, 90–, 80–, 70–, 60–, 50–, 40–, 30–, 20–, 10–, 0–0

410 — DEGREE OF URBANITY
Standard metropolitan area population expressed in millions.

Scale: 100–2, 90–, 80–, 70–, 60–, 50–, 40–, 30–, 20–, 10–, 0–0

420 — PROXIMITY TO RELATIVES
One-way air travel time for visiting relatives expressed in hours.

Scale: 100–0, 90–, 80–, 70–, 60–, 50–, 40–, 30–, 20–, 10–, 0–5

Figure 27.   Elementary criteria for job selection

198

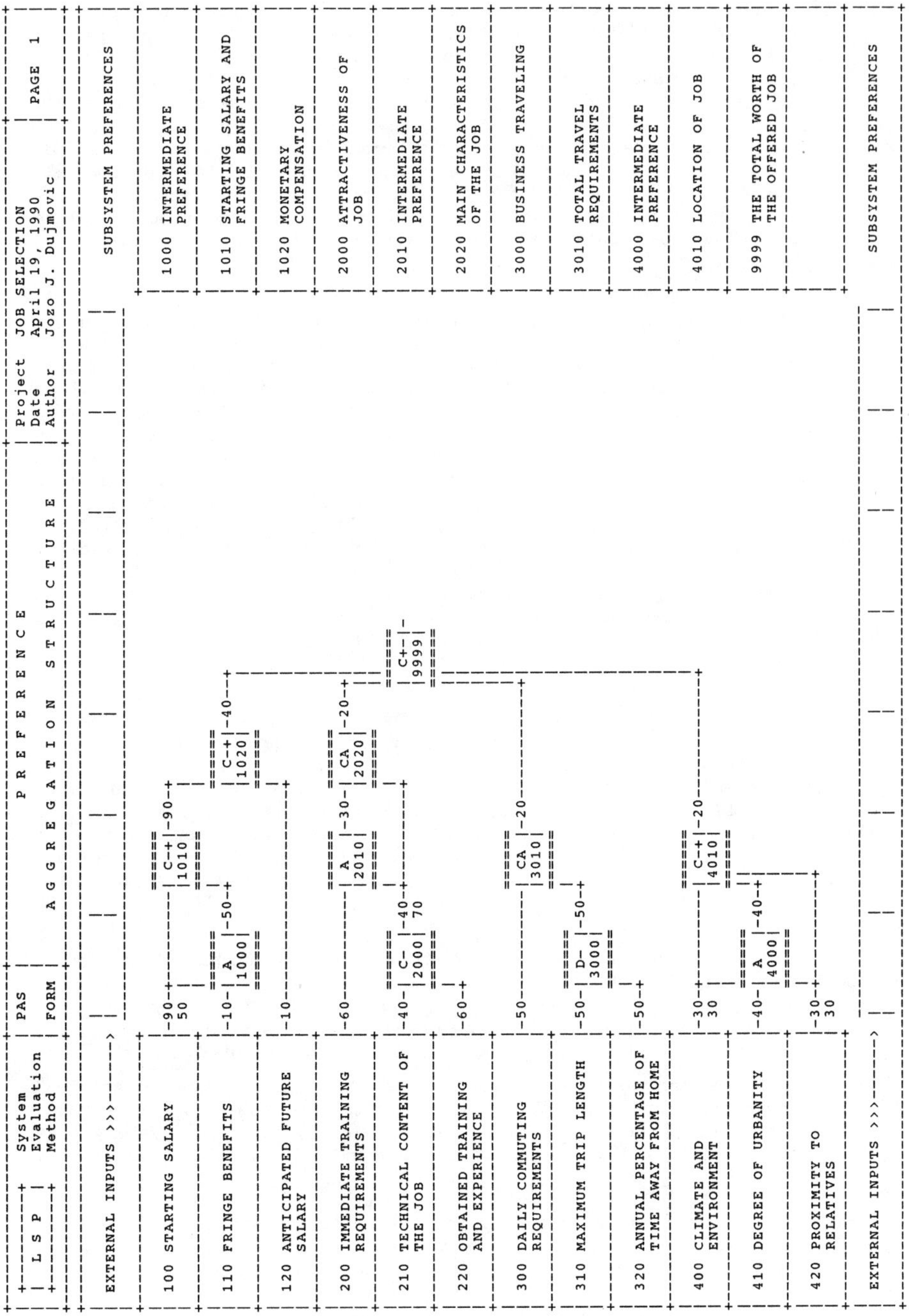

Figure 28.   PNN for job selection

199

definitions of relative performance variables. For example, all monetary compensations are evaluated so that the maximum offered compensation is assigned the maximum preference 100%, and a compensation that is 50% of the maximum offered compensation is assigned the preference 0. Finally, for some performance variables we cannot define exact preference scales and must rely on a direct preference assessment by the evaluator. For example, the technical content of the job (elementary criterion #210) may be considered important for a sound decision but we cannot propose a scalar variable that objectively measures the analyzed technical content. Such elementary preferences can only be directly assessed by the evaluator. The direct preference assessment, as well as all adjustable parameters of criterion function, fully depend on evaluator's expertise.

The PNN for aggregation of preferences is presented in Fig. 28. Some of the inputs are considered mandatory requirements, and those include the starting salary, the anticipated future salary, daily commuting requirements, climate and environment, and the proximity to relatives. If any of these inputs is rated zero, then the job offer as a whole will also be rated zero (the input preferences propagate through neurons that use the compensation degree below the low compensation threshold value). In addition, some subsystem preferences are also treated as mandatory (e.g. business traveling and the attractiveness of the job). The presented PNN includes three examples of the use of PA-ADAPRENE's. First, the starting salary and the fringe benefits are not logically equivalent: a good starting salary is considered necessary, while the fringe benefits are desired, but not necessary. Similarly, the low immediate training requirements are desired, while the attractiveness of the job, expressed by the criteria #210 and #220, is considered necessary. The degree of urbanity is defined as desired and the remaining components of the location of the job are necessary.

Figures 27 and 28 are generated by the Criterion Development System (CDS) [DUJ76a], a software tool that is used for the development of complex criteria and preparation of their documentation. The modeling of preferential neurons, based on a numerical technique for computing weighted power means developed in [SLA74], is performed by programs written in a System Evaluation Language (SEL) [DUJ76b]. In addition to system evaluation (i.e. the computation of preferences through the layers of PNN) SEL also includes instructions for cost-preference analysis, sensitivity analysis, and system optimization based on the optimization technique proposed in [DUJ74c].

The cost-preference analysis [DUJ82, DUJ87] is a technique for selecting the best system in a similar way as in the case of classic cost-effectiveness [SEI69, OST77, SAG77]. The technique can be applied for selecting computers and similar technical systems where each of competitive systems is assigned a global preference E_0 and a global cost C_0 (e.g. the presented job selection model does not belong to this category). The global preference is computed from the PNN model, and the global cost is obtained from a suitable cost analysis (see [DUJ82] and [SU87]). We first select only those systems that satisfy the conditions $E_0 \geq E_{min}$ and $C_0 \leq C_{max}$, where E_{min} denotes the minimum acceptable global preference (typically, $E_{min} = 67\%$) and C_{max} denotes the maximum cost the user is ready to pay for the evaluated system. The selected systems can then be compared according to various global quality indicators that are defined as functions of E_0 and C_0. If the maximum preference achieved by competitive systems is E_{max} and the minimum achieved cost is C_{min}, then some of global quality indicators (Q) can be defined as follows:

$$Q_u := E_0/C_0 \ ,$$

$$Q_r := pE_0/E_{max} + (1-p)C_{min}/C_0 \ , \qquad 0<p<1 \ ,$$

$$Q_e := pE_0 + (1-p)(C_{max}-C_0)/C_{max} \ ,$$

$$Q_c := pC_{min}E_{max}/E_0 + (1-p)C_0$$

The global quality indicator Q_u shows the relative degree of satisfaction of requirements per monetary unit of total investment, while Q_r, Q_e, and Q_c combine cost and preference using the weight p for preference and the weight $1-p$ for cost. Q_r is a dimensionless indicator, Q_e is an effective global preference, and Q_c is an effective global cost. The best system can be selected according to the maximum value of Q_u, or Q_r, or Q_e, or according to the minimum value of Q_c.

The system optimization using PNN's is a technique that can be used in cases where the performance variables x_1, x_2, ..., x_n are related to corresponding elementary costs c_1, c_2, ..., c_n. For example, x_i may be the capacity of disk memory and c_i is the corresponding cost of disk memory. Assuming that we know the functions $x_i = z_i(c_i)$, $i=1,...,n$, the global preference of a system can be written as follows:

$$E_0 = G(x_1, x_2, \ldots, x_n) = G(z_1(c_1), z_2(c_2), \ldots, z_n(c_n)) = U(c_1, c_2, \ldots, c_n)$$

Now we can formulate the following two questions:

(1) *What is the optimum system configuration yielding the maximum global preference for a given constrained cost of the system?*

(2) *What is the minimum cost of a system necessary to attain a given level of global preference (i.e. to satisfy a given percentage of user's requirements)?*

A solution of these problems, proposed in [DUJ74c] and implemented as SEL instructions *OPTIMIZE* AND *ALLOCATE* [DUJ76b], yields the optimum preference

$$E^{opt} = \max_{c_1+\ldots+c_n \leq C} U(c_1, \ldots, c_n) = U(c_1^{opt}, \ldots, c_n^{opt}) = G(x_1^{opt}, \ldots, x_n^{opt})$$

$$= A^{opt}(C) \ , \qquad 0<C\leq C_{max} \ .$$

The optimum allocation of resources is defined by the allocation functions a_i, $i=1,\ldots,n$:

$$c_i^{opt} = a_i(C) \ , \quad i=1,\ldots,n \ ,$$

and the corresponding optimum system configuration parameters are:

$$x_i^{opt} = z_i(c_i^{opt}) = z_i(a_i(C)) \ , \quad i=1,\ldots,n \ .$$

These results answer the above two questions.

The most frequent applications of preferential neural networks were in the area of evaluation and selection of digital computers [DUJ75c, DUJ77a, DUJ80b, DUJ87], and so far more than 20 projects of medium and large computer evaluation, comparison and selection, for major governmental and industrial organizations have been successfully completed. In comparison with traditional computer selection methods [SHA69, TIM73] that are usually based on qualitative analyses [MCQ78], various cost-value and utility oriented quantitative techniques [JOS68, KLE80], or simplistic scoring approaches [GIL71, GIL76, DUJ77b, KLE78], the preferential neural networks offer three important advantages: (1) PNN's are based on continuous preference logic and offer high flexibility in expressing any kind of complex logical relationship among system requirements, (2) the stepwise supervised learning technique enables an easy and extensive acquisition of expert experience that yields a high accuracy of PNN models, and (3) the available specialized software tools (ANSY, CDS, and SEL) greatly reduce the time necessary to organize and efficiently use the PNN's. In addition to digital computer evaluation and selection studies, the PNN models were also successfully used for the evaluation and optimization of analog computers [DUJ74d, DUJ76c], hybrid computers [DUJ76d], and data management systems [SU87].

10. CONCLUSION

We have presented a survey of continuous preference logic and its application in the area of system evaluation models. The CPL was developed to help us understand the mental process of system evaluation, and to serve as a mathematical background for deriving quantitative system evaluation models. The system evaluation models can be efficiently organized as preferential neural networks. The processing units of such networks are preferential neurons that implement various CPL connectives. The fundamental CPL connective is the generalized conjunction-disjunction and we have suggested a set of 12 conditions that the GCD should satisfy in order to be suitable as a main component for the realization of preferential neurons.

The training of preferential neural networks is a rather specific problem. The preferential neural networks describe and model a mental process and consequently they must reflect the expert knowledge and evaluation criteria of human decision makers. So, the preferential neural networks for system evaluation cannot be derived from objective reality in the way we can derive traditional neural networks for pattern recognition or signal processing. Preferential neural networks are a rare example where the classic training set for supervised learning cannot be available. The number of inputs is too large and human decision makers cannot specify reliable data for huge training sets. Fortunately, preferential neural networks for system evaluation can be systematically organized as tree-like structures consisting of a small number of basic neurons. We have used that specific structure of the PNN to propose a stepwise training technique where instead of training the whole network the decision maker separately trains each preferential neuron. The resulting neural network can then be used for evaluation, comparison, selection, and optimization of complex systems. Future research should show whether preferential neural networks can be used for solving problems in other areas of traditional neural network applications.

REFERENCES

CSS89   California Scientific Software: Brain Maker - Users Guide and Reference Manual. 3rd Edition, January, 1989.

COU77   Coutinho, J. de S., Advanced Systems Development Management. John Wiley and Sons, 1977.

DOM82a   Dombi, J. A general class of fuzzy operators, the De Morgan class of fuzzy operators and fuzziness measures induced by fuzzy operators. Fuzzy Sets and Systems 8 (1982) pp. 149-163.

DOM82b   Dombi, J., Basic concepts for a theory of evaluation: the aggregative operator. Europ. J. Operat. Res. 10 (1982) pp. 282-293.

DUB85   Dubois, D. and H. Prade, A review of Fuzzy Set Aggregation Connectives. Information Sciences 36 (1985), pp. 85-121.

DUJ73a   Dujmović, J.J., Two Integrals Related to Means. Journal of the University of Belgrade EE Dept., Series Mathematics and Physics, No. 412 - No. 460, 1973, pp. 231-232.

DUJ73b   Dujmović, J.J., A Generalization of Some Functions in Continuous Mathematical Logic - the Evaluation Function and its Applications. (In Serbo-Croatian). Proceedings of the Informatica Congress, Bled, Yugoslavia, 1973, paper D27.

DUJ73c   Dujmović, J.J., Mixed Averaging by Levels (MAL) - a System and Computer Evaluation Method, (In Serbo-Croatian). Proceedings of the Informatica Congress, Bled, Yugoslavia, 1973, paper D28.

DUJ74a   Dujmović, J.J., Weighted Conjunctive and Disjunctive Means and their Application in System Evaluation. Journal of the University of Belgrade, EE Dept., Series Mathematics and Physics, No. 483, 1974, pp. 147-158.

DUJ74b   Dujmović, J.J., New Results in Development of the "Mixed Averaging by Levels" Method for System Evaluation. (In Serbo-Croatian). Proceedings of the Informatica Congress, Bled, Yugoslavia, 1974.

DUJ74c   Dujmović, J.J., Optimization of Complex Systems Using the MAL Method. (In Serbo-Croatian). Proceedings of the Informatica Congress, Bled, Yugoslavia, 1974.

DUJ74d   Dujmović, J.J., and Dzigurski, O.D., Evaluation and Comparison of Analog Computers. (In Serbo-Croatian). Proceedings of the Informatica Congress, Bled, Yugoslavia, 1974.

DUJ75a   Dujmović J.J., A Graphic Approach to Weighted Conjunctive and Disjunctive Means Calculation. Journal of the University of Belgrade. EE Dept., Series Mathematics and Physics, No. 536, 1975, pp. 191-196.

DUJ75b   Dujmović, J.J., Extended Continuous Logic and the Theory of Complex Criteria. Journal of the University of Belgrade, EE Dept., Series Mathematics and Physics, No. 537, 1975, pp. 197-216.

DUJ75c   Dujmović, J.J., Evaluation of Digital Computers Using the System Evaluation Method MAL, (In Serbo-Croatian). Proceedings of the

Informatica Congress, Bled, Yugoslavia, 1975.

DUJ76a   Dujmović, J.J., and I. Tomašević, Criterion Development System (CDS), (in Serbo-Croatian). Proceedings of the Informatica Congress, Bled, Yugoslavia, 1976, paper 1/122.

DUJ76b   Dujmović, J.J., System Evaluation Language (SEL) - a Programming Language for Evaluation, Comparison and Optimization of Complex Systems, (In Serbo-Croatian). Proceedings of the Informatica Congress, Bled, Yugoslavia, 1976, Paper 1/121.

DUJ76c   Dujmović, J.J., A Technique for Determining Optimal Configurations of the Computing Units of Analog and Hybrid Computers, (In Serbo-Croatian). Proceedings of the ETAN Conference, 1976, pp. 1181-1188.

DUJ76d   Dujmović, J.J., Evaluation, Comparison and Optimization of Hybrid Computers Using the Theory of Complex Criteria. Simulation of Systems. Dekker L. (Ed.). North-Holland, Amsterdam 1976, pp. 553-566.

DUJ77a   Dujmović, J. J., "Professional Evaluation and Selection of Computer Systems," Proceedings of the 12th Yugoslav International Symposium on Information Processing, Paper 4-001, October 1977.

DUJ77b   Dujmović, J.J., The Preference Scoring Method for Decision Making - Survey, Classification, and Annotated Bibliography. Informatica, No. 2, 1977, pp. 26-34.

DUJ80a   Dujmović, J.J., Partial Absorption Function. Journal of the University of Belgrade, EE Dept., Series Mathematics and Physics, No. 659, 1980, pp. 156-163.

DUJ80b   Dujmović, J.J., Computer Selection and Criteria for Computer Performance Evaluation. International Journal of Computer and Information Sciences, Vol. 9, No. 6, Dec. 1980, pp. 435-458.

DUJ82   Dujmović, J.J. and R. Elnicki, A DMS Cost/Benefit Decision Model: Mathematical Models for Data Management System Evaluation, Comparison, and Selection. National Bureau of Standards, Washington D.C., No. GCR 82-374. NTIS No. PB 82-170150 (150 pp.), 1982.

DUJ87   Dujmović, J.J., The LSP Method for Evaluation and Selection of Computer and Communication Equipment. Proceedings of MELECON'87, Mediterranean Electrotechnical Conference and 34th Congress on Electronics (Joint Conference), IEEE/RIENA, Rome, Italy, 1987, pp. 251-254.

DUJ86   Dujmović, K.A., Classification of Objects Using Preference Logic Functions (In Serbo-Croatian). MS Thesis, Center for Multidisciplinary Studies, University of Belgrade, 1986.

DYC84   Dyckhoff, H. and W. Pedrycz, Generalized means as model of compensative connectives. Fuzzy Sets and Systems 14 (1984) pp. 143-154.

DYC85   Dyckhoff, H., Basic concepts for a theory of evaluation: Hierarchical aggregation via autodistributive connectives in fuzzy set theory. Europ. J. Operat. Res. 20 (1985) pp. 221-233.

FIS64   Fishburn, P. C., Decision and Value Theory, J. Wiley, 1964.

FIS70   Fishburn, P. C., Utility Theory for Decision Making, J. Wiley, 1970.

GIL71   Gilb, T., Reliable Data Systems. Universitetsforlaget, Oslo, 1971.

GIL76   Gilb, T., Software Metrics. Studentlitteratur, Lund, 1976.

GIN58   Gini, C. et al. Le Medie. Unione Tipografico-Editrice Torinese. Milano, 1958. (Russian translation published by Statistika, Moscow, 1970)

HEC87   Hech-Nielsen, R., Kolmogorov's Mapping Neural Network Existence Theorem. Proceedings of the IEEE First International Conference on Neural Networks, Edited by M. Caudill and C. Butler, San Diego, 1987, pp. III-11 - III-13.

HIL69   Hillegass, J. R., "Systematic Techniques for Computer Evaluation and Selection," Management Sci., July-August 1969, pp. 36-40.

JOH77   Johnson, E. M. and G. P. Huber, "The Technology of Utility Assessment," IEEE Trans. Syst., Man, Cybern., Vol. SMC-7, May 1977, pp. 311-325.

JOS68   Joslin, E. O., "Computer Selection," Addison-Wesley Publishing Company, 1968. (An augmented edition was published in 1977 by The Technology Press Inc., Box 125, Fairfax Station, Virginia 22039.)

KEE76   Keeney R. L. and H. Raiffa, "Decisions with Multiple Objectives/ Preferences and Value Tradeoffs," J. Wiley, 1976.

KLE78   Kleijnen, J. P. C., "Scoring Methods, Multiple Criteria, and Utility Analysis," Research Memorandum, Tilburg University, Holland, Department of Economics, October 1978. (Published also in Sigmetrics "Performance Evaluation Review".)

KLE80   Kleijnen, J. P. C., "Computers and Profits," Addison-Wesley, 1980.

MAC73   MacCrimmon, K. R., "An Overview of Multiple Objective Decision Making," in "Multiple Criteria Decision Making," J. L. Cochraine and M. Zeleny, Editors., University of South Carolina Press, pp. 18-44, 1973.

MCQ78   McQuaker, R. J., "Computer Choice," North-Holland, Amsterdam, 1978.

MIL66   Miller, J. R., III, "The Assessment of Worth/a Systematic Procedure and its Experimental Validation," Ph.D. Dissertation, M.I.T., 1966.

MIL70   Miller, J. R., III, "Professional Decision-Making," Praeger, New York, 1970.

MIT69   Mitrinović, D.S. and P.M. Vasić, "Means" (In Serbo-Croatian), Mathematical Library, Vol. 40, Belgrade, 1969.

MIT77   Mitrinović, D.S., P.S. Bullen and P.M. Vasić, "Means and Related Inequalities" (In Serbo-Croatian), Publications of the University of Belgrade EE Dept., No. 600, Belgrade, 1977.

NEL64   Nelder, J.A., and R. Mead, A Simplex Method for Function Minimization. Computer Journal, Vol. 7, No. 4, 1964, pp. 308-313.

OST77   Ostrofski, B., "Design, Planning and Development Methodology," Prentice-Hall, 1977.

SAG77   Sage, A. P., "Methodology for Large-Scale Systems," McGraw-Hill Book Co., 1977.

SAL83   Salton, G., E. A. Fox, and H. Wu, Extended Boolean Information Retrieval. Com. of the ACM, Vol. 26, No. 12, 1983, pp. 1022-1036.

SEI69   Seiler III, K., "Introduction to Systems Cost-Effectiveness," Wiley-Interscience, 1969.

SHA69   Sharpe, W. F., "The Economics of Computers," A RAND Corporation Research Study, Columbia University Press, New York, 1969.

SLA74   Slavić, D. V. and Dujmović, J. J., Numerical Computation of Weighted Power Means. Journal of the University of Belgrade, EE Dept., Series Mathematics and Physics, No. 485, 1974, pp. 167-171.

SOU88   Souček, B. and M. Souček, Neural and Massively Parallel Computers. J. Wiley & Sons, 1988.

SU87   Su, S. Y. W., Dujmović, J. J., Batory, D. S., Navathe, S. B., and R. Elnicki, A Cost-Benefit Decision Model: Analysis, Comparison, and Selection of Data Management Systems. ACM Transactions on Database Systems, Vol. 12, No. 3, September 1987, pp. 472-520.

TIM73   Timreck, E. M., "Computer Selection Methodology," Computing Surveys, Vol. 5, No. 4, 1973, pp. 199-222.

WHI75   White, D. J., Decision Methodology. John Wiley & Sons, 1975.

WHI63   White, D. R. J., D. L. Scott, and R. N. Schulz, "POED - A Method of Evaluating System Performance," IEEE TEM, December 1963.

YAG80   Yager, R. R., On a general class of fuzzy connectives. Fuzzy Sets and Systems 4 (1980) pp. 235-242.

YAG87a Yager, R. R., On the aggregation of processing units in neural networks. Proceedings of the IEEE First International Conference on Neural Networks, Edited by M. Caudill and C. Butler, San Diego, 1987, pp. II-327 - II-333

YAG87b Yager, R. R., A Note on Weighted Queries in Information Retrieval Systems. J. of the American Society for Information Science 38(1), 1987, pp. 23-24.

ZIM80   Zimmermann, H. J. and P. Zysno, Latent connectives in human decision making. Fuzzy Sets and Systems 4 (1980) pp. 37-51.

ZIM83a Zimmermann, H. J., Using fuzzy sets in operational research. European Journal of Operational Research 13 (1983) pp. 201-216.

ZIM83b Zimmermann, H. J., and P. Zysno, Decisions and evaluations by hierarchical aggregation of information. Fuzzy Sets and Systems 10 (1983) pp. 243-260.

8

ANALOG VLSI SYSTEMS
FOR NEURAL ADAPTIVE ARCHITECTURES

Daniele D. Caviglia, Maurizio Valle,
and Giacomo M. Bisio

Department of Biophysical and Electronic Engineering
University of Genoa - via Opera Pia 11/a
I-16145 Genoa - Italy

0. Abstract

The chapter describes a VLSI-compatible analog adaptive multilayer architecture for the Back Propagation (BP) algorithm.

The synaptic matrix is fully programmable. The discretized values of weights are controlled by analog voltages dynamically restored on capacitors at each synaptic site. The weight changes are determined by a combination of information local to each weight and signals generated by each neuron and back-propagated through the rows of the synaptic matrix.

The effect of weight discretization on the functionality of the BP algorithm is considered both for a linear and an exponential dependence of weights with respect to the controlling voltage.

A complete circuit implementation of the adaptive architecture is presented and analyzed through simulation.

1. Introduction

1.1 Neural Computational Models

Neural Networks (NN's) are a computing paradigm peculiar with respect to both the parallel computational elements upon which it is based and the way their functionality is programmed ($\Rightarrow$ *learning*).

The reasons for the current vast interest in neural network models are numerous, reflecting the multidisciplinary nature of the subject. For electrical engineers and computer scientists a major stimulus has been the development of VLSI computational systems of new conception based on neural architectures.

The modelling of neural computation has been pursued following two different directions.

The first, in its conceptual essence, can be traced back to the Mc Culloch-Pitt's neuron model [McPi43]. The neuron is performing simple threshold logic, and these simple elements connected in a network can implement any finite logical expression. Along this path, *connessionism* emphasizes connection complexity and processor simplicity (e.g. 0-1 threshold detector [RuMc86]).

The second direction is the attempt to create (biologically motivated) structured neural systems that perform (specific) computations with high efficiency. The computational activity generally is not a logic funtion, but a sort of *spatially distributed analog computation* ,[AnRo88,PiMc47]. This approach has been denoted Biophysics of Computation by Koch and Poggio [KoPo84]. Their analysis evidences that (1) the complexity of the processing that takes place within a single neuron may be far greater than simple thresholding; (2) several different mechanisms of information processing are present in biological systems; (3) analog processing is present and relevant for performance.

In this context one of the most interesting contributions is the silicon implementation of a retina model by Mead et al [MeMa88].

1.2 Electronic Neural Architectures

When one examines the peculiarities of present VLSI technology (2+1/2 dimensions of circuits, speed, programmability, pipelining, etc.), looking for VLSI implementations of neural networks, two main directions can be distinguished:

First, development of parallel VLSI architectures, based on non-standard microprocesors (e.g. transputers), to efficiently simulate neural nets. These studies provide quantitative insight into performances of both neural algorithms and their VLSI implementation at architectural level [Ca/88, IEMI89]. Massively parallel architectures based on custom chips for the emulation of neural systems can also be considered.

Second, design of VLSI "neural hardware" that emulates biological neural architectures [IEMI89, Mea89, MeIs89]. These implementations do not rely on standard circuitry, but adopt innovative analog and digital architectures, eventually employing specific technological variants (high resistivity compounds, weak-inversion driven MOS transistors, etc.). These studies lead to the development of special purpose modules to be operated in connection with a more conventional host computer.

1.3 Implementation Medium: MOS technology

In contrast to bipolar technology, MOS integrated circuits offer the ability to store charge on a node over a period of even seconds, and to sense the value of the charge continuosly and non-destructively. The former results from the high impedance of the MOS transistor in the off state, and the latter results from the essentially infinite input impedance of the device.

The information stored on nodal capacitances is essentially analog: the node voltage has a continuous range of values. In digital applications the information is usually interpreted as being only one of *two* logic levels; these levels are typically refreshed (restored) every few milliseconds. However various examples of finer discretization have been reported in the literature [Ao/87],[Ho/88],[Hoc89] to achieve a higher storage capacity.

Extending this approach, one could approximate an analog memory with a sequence of discrete levels, thus gaining in programmability and avoiding the problems to be faced in pursuing purely analog techniques [Ho/90].

Various other solutions have been contemplated for analog weight storage, depending on the technology available and the peculiarities of the storage function to be implemented [Bo/89, Sc/89,Vit90].

Some computations require the continuous updating of synaptic weight, but do not need long time storage. Careful design and cooling the chip can reduce leakage currents to a level low enough to use analog MOS storage without refresh [Sc/89].

Another approach resorts to EEPROM technology, using isolated floating-gate structures to store the weight values [Bo/89].

Another alternative, more conventional, uses digital storage with analog to digital and digital to analog conversions.

A further consideration is that to integrate a large number of analog neural circuits on a single chip it is mandatory to have low power dissipation. This can be achieved by operating the MOS transistors in the sub-threshold region [Vit85],[Mea89]. In this region the dependence of the drain current on gate voltage is exponential, thus providing through a physical mechanism a functional behavior useful for the algorithm to be implemented. This will require to accurately assess the functionality of neural algorithms and particularly of their learning capabilities in relation to the analog hardware implementation chosen.

1.4   Summary

In this chapter we address two main issues: i) the formulation of a modified BP algorithm based on discrete weights; ii) the definition of an analog target architecture and the implementation of its basic elements, characterized by various degrees of programmability .

2.   An Analog Neural Architecture for the Back Propagation Algorithm

Computer architectures, in terms of instruction set and storage models, are defined through objects, actions and capabilities presumed to be present in a physical host computer. Therefore our investigation begins with a comparative analysis of (1) the neural computational task to be performed both in the input processing and in the learning phases; (2) the specific capabilities offered by analog implementations.

2.1 Back Propagation Algorithm for Discretized Weights

We have chosen to consider the computational task of Back Propagation (BP) algorithm since multilayer perceptrons, trained with the basic BP algorithm have been used successfully as pattern classifier in many applications [RuMc86], even though examples have been found disproving the presumption that, barring local minima, BP will find the best set of weights for a given problem [Br/89].

The hardware implementation of learning neural networks requires the availability of a proper circuit technique for varying and maintaining the weights of synaptic connections. Not being conceivable a truly analog memory, it is necessary to introduce discretized values for the weights. Therefore the effects of the discretization on the functionality of the neural algorithm should also be assessed.

We first analyse the modifications of BP algorith proposed by T.P. Vogl et al. [Vo/88] for accelerating its convergence (see Box 1).

THE VOGL's ALGORITHM [Vo/88]

The Vogl's variant of BP algorithm consists on three main points:

(1) The weight modification algorithm is operated by "epoch", i.e. the network weights are not updated after each pattern is presented; the changes of each weight are summed over all the input patterns (p), $\sum_p \delta_{pj} O_{pi}$, and the sum is applied to modify the weight only once per iteration (m) through all the patterns:

$$\Delta W_{ji}(m+1) = \eta \cdot \sum_p \delta_{pj} O_{pi} + \alpha \Delta W_{ji}(m)$$

where W_{ji} is the weight of the synapse from the i-th input to the j-th output neuron, η is the learning rate (i.e. the multiplier on the step size), O_{pi} is the output of neuron i at the p-th pattern presentation, δ_{pj} is a measure of the error for neuron j, and α is the momentum factor (i.e. a measure of the "memory" of previous steps).

(2) The learning rate η is varied according to whether or not an iteration decreases the total error for all patterns. If an update results in a reduced total error, η is multiplied by a factor $\phi > 1$ for the next iteration. If an iteration step produces a network with a total error larger than the previous value more than a few percent, the weight changes are rejected, η is multiplied by a factor $\beta < 1$, and

(3) α is set to zero and the step is repeated till a successful step occurs: in this case α is reset to its initial value.

The stopping criterion can be stated in various ways: typically, the learning iterations are repeated until $\max_{p,j} | t_{pj} - O_{pj} | \leq \varepsilon$, where t_{pj} is the j-th value of the p-th target pattern, and ε is the maximum error accepted.

Box 1

Considering the selection of weights as a problem in non-linear optimization theory, the rationale behind these modifications is : i) to avoid misdirection in the optimization path; ii) to adapt the step size to the local shape of the hyper-dimensional surface being traversed in the optimization; iii) to escape from shallow local minima by accepting steps in the optimization path producing an increase in the total error of a few percent.

The discretization of the weights can deepen the local minima already present and also introduces many spurious local minima, not corresponding to any meaningful solution. To avoid remaining trapped in these minima proper modifications of the stepping parameters of the Vogl's algorithm has been proposed in [Ca/90]. The new algorithm is described in Box 2.

2.1.1 A classification problem with discretized BP algorithm

To assess the performances of these algorithms with discrete weights the following classification problem has been studied through simulations [Ca/90]: a 2-layer network with 35 inputs, 18 hidden units, 10 output neurons has been used to classify typewritten characters (the digits 0-9), coded by using a 7×5 pixel matrix (see Fig. 1a).

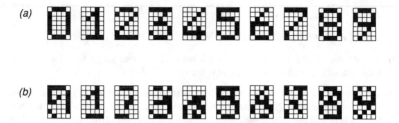

(a)

(b)

Fig. 1. In the first row (a) the original patterns for the 10 digits are shown. In the second row (b) is reported one set of noisy patterns obtained with a 15% error probability.

The simulation with discretized weights is performed according to the algorithm in Box 2 (MV = Modified Vogl's). The algorithm parameters are reported in Table I in the PD column.

Two different discretization strategies have been followed: the first with steps of constant amplitude ΔW ; the second one with steps that maintain constant the maximum relative discretization error $\frac{\Delta W}{W}$. The latter implies an exponential increase in the step size. For a given number of steps, the second approach allows to obtain a better resolution when it is needed - when the synapses strength is small - and a poorer when it is acceptable [VNeu58].

Three different values in the number of levels are considered (16, 24, 32) for both strategies, and the discretization is performed at the end of each epoch, after the weights has been updated .

	Original parameters (PA)	Modified parameters (PD)
α_{start}	0.9	0
η_{start}	0.1	0.1
Φ	1.05	1.2
β	0.7	0.7
ε	0.05	0.2

Table I. Parameters for the different discretization techniques. In the PA column the original Vogl's values are reported; in the PD column is shown the set of values used for the case of discrete weights.

These simulations have been compared: (i) with the standard BP algorithm by epoch (ES in the first row); (ii) with two fully analog situations using the same PA and PD sets of parameters (first and second columns).

The results presented in the first three rows of Table II refer to an epoch of 10 patterns not corrupted by noise. The MV algorithm has been also tested with noisy patterns (last row in Table II): each pixel has a 15% probability of assuming a wrong value (see Fig. 1b). In this case the epoch spans over 100 patterns.

The number of iterations was allowed to reach 10.000. In the discretized cases the iterations normally stop when the output O_{pj} of the neuron j in correspondence to the input pattern p is higher than 0.7 (being the target value $t_{pj} = 0.9$ and $\varepsilon = 0.2$), and all the other outputs $O_{pi} \mid i \neq j$ are lower than 0.3 (being $t_{pi} = 0.1$).

The reasons for having failed to reach the convergence after 10.000 steps have been examined determining the percentage of patterns that did not satisfy the error conditions. For instance, the case MV-noisy patterns with 32 discrete levels exponentially spaced fails to reach convergence one time over ten attempts. But only one over 100 patterns is responsible of this failure. In particular, with reference to Fig. 1b, it occurred for the character "3": the output of the "3" neuron was 0.76 (which is correct), while the output of the "5" neuron was 0.31, which does not satisfy the condition on the maximum allowed error.

The table shows how the exponential discretization approach provides better performances with respect to the linear case.

	Analog		Discrete					
	PD	PA	C32	C24	C16	E32	E24	E16
ES	10%	10%	100%	100%	100%	100%	100%	100%
	10%	10%	100%	100%	100%	100%	100%	100%
	2318	2320	-	-	-	-	-	-
MV	0%	0%	10%	20%	70%	0%	0%	20%
	-	-	10%	15%	30%	-	-	25%
	249	528	433	526	796	382	471	706
MV - noisy patterns	2%	0%	0%	6%	39%	10%	20%	60%
	10%	-	-	34%	56%	1%	20%	21%
	3722	5844	5160	7360	5979	6020	5250	8550

TABLE II: Results of simulations. The numbers in each box indicate respectively, top to bottom: percentage of simulations that did not converge after 10.000 iterations; percentage of patterns that, in those cases, did not satisfy the error conditions; average number of iterations when classification is successful.

2.2 The Learning Architecture

The architecture considered here is organized in the form of a Multilayer Perceptron to be trained with the BP algorithm. Each layer consists of a regular matrix of synapses connected to an array of neurons in a crossbar arrangement (Fig. 2). Each layer performs both the feed-forward computations and the back propagating updates of weights for learning. The functionalities of neurons and synapses are more complex than what usually occurs. The weights of synapses are controlled by a mechanism simple enough to be reproduced at each weight site. The weight changes are determined by a combination of informations local to each weight and signals generated by each neuron and back-propagated through the array in a row-wise fashion.

Two main blocks describe the functionality of the neuron. The operational amplifier f responds with a sigmoidal shape to the sum of input synaptic currents; block f' computes an approximation of the derivative of the output function of f using the

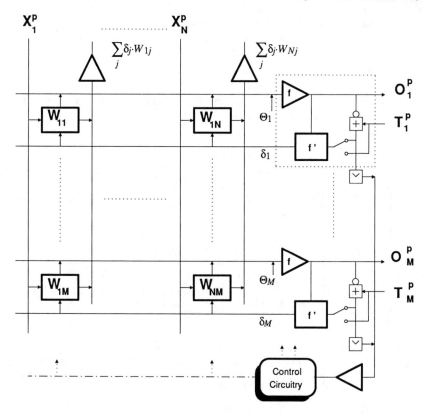

Fig. 2. Analog Architecture for the modified BP algorithm.

214

analog product of f and (1-f) and multiplies it by a measure of the error to obtain the factor δ_j. The output section of the neurons generates also the error signal for the Control Logic. This block, besides generating the synchronizations for all the circuits, takes into account the variations of the error and changes the factor η according with the modified Vogl's algorithm, till convergence occurs in the learning procedure.

Fig. 3 shows the basic building blocks of the adaptive synapses. Block F performs the feed-forward computation: the output current I_{ij}, is proportional to the product of the input voltage X_i times the transconductance W_{ij}. Block B1 back propagates the error ($\delta_j \cdot W_{ij}$) toward the preceeding layer. Block B2 generates the weight update ΔW_{ij} ; in the case of a learning procedure "by pattern" the B2 multiplier generates the update directly; if the learning occurs "by epoch" the weight changes are accumulated integrating current pulses on a capacitor.

The circuitry devoted to the weight refresh is not shown in Fig. 2 for sake of clarity, and also because it will not modify the architecture. Implementation details are reported in the following paragraph. However, an important question has to be faced here concerning the synchronization of the learning phase with the refresh phase. It can be noted that if the learning algorithm operates "by pattern" it results in a continuous updating of weights, thus compensating the decay due to leakage through the adaptability of the algorithm. In this case there is no need of refresh until the learning is complete. In the case of learning "by epoch" the problem depends on how many patterns the epoch consists of. Also in most of these cases it is not necessary to operate in the refresh mode during learning. However, with large sets of patterns (on the order of 10^4), we can introduce a refreshing phase (and consequenly a discretization) several times during an epoch, as in the cases discussed in Section 2.1. The effect of the decay of the charge on the capacitors used to store ΔW_{ij} can be counteracted by varying the order of presentation of the patterns.

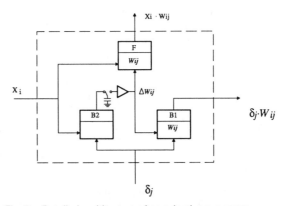

Fig. 3. Detailed architecture of an adaptive synapses.

A final consideration has to be made about the pin count of a chip implementing this architecture. The pins carrying input and output signals during the feedforward computation, can be multiplexed in time to back propagate the error signals for the computation of the weight changes during learning, thus reducing the total pin number.

3. Analog CMOS Implementation

The functionality of the various components considered in the previous paragraph is related to their circuit implementation , developed following the design principles proposed in [Mea89, Vit90]. The basic operators used in the network consist in various types of multipliers, adders, storage cells, absolute-value circuits (rectifiers), amplifiers, in addition to traditional digital control logic for the synchronization with the external circuitry. Following [Mea89] we employ standard CMOS components to implement these blocks.

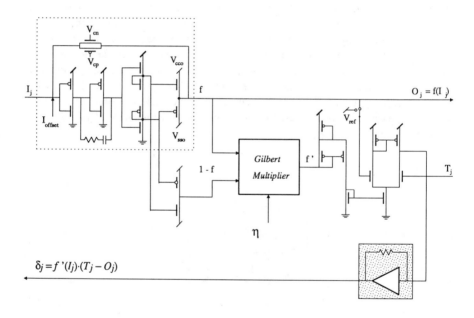

Fig. 4. *Circuit for the neuron. The shaded block is a simple amplifier similar to block f. The position for the switch corresponds to a neuron for an output layer; the other position is to be used for hidden layers.*

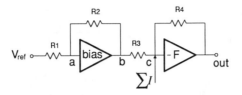

Fig. 5 . Bias circuit for neuron F. If the amplifiers operate in their linear region the voltages at nodes a and c are equal but different from V_{ref} ; if all the resistances are equal, the voltage at node out is equal to V_{ref} within a few millivolts.

Due to the use of nonlinearity and adaptation, some of the typical analog design requirements (accurate component values, device matching, precise time constants, etc.) are relaxed somewhat when implementing neural networks [BiIs90].

3.1 Neuron circuit

The circuit in the dotted box of Fig. 4 implements the neuron op-amp f: the algebraic sums of synaptic currents is applied to the virtual ground obtained by the feedback at the input of the first inverter.

The two inverters constitute the gain stage that is followed by an inverting complementary voltage shifter and the output buffer. With this configuration the circuit operates as a class AB amplifier with low quiescent current; consequenlty the static power dissipation of the output transistors is very low.

The voltage at the input node depends on the device parameters and is approximately midway from ground to V_{dd} ; to hold the output voltage at V_{ref} when the total current in input $I_j = 0$ it is necessary to add a compensating current I_{offset} with a bias circuit (see Fig. 5).

Since the outputs of neurons of the first layer are connected to the inputs of the second layer synapses, the corresponding voltage values must fit to the operating range of the weak inversion OTA. This means that low power supply is to be adopted for the output stage of the neuron (e.g. $V_{sso} = 2.5$ V, $V_{cco} = 2.6$ V).

The complementary MOS transfer gate on the top acts as the feedback resistor that determines the closed loop gain.

The Figure 4 shows also the circuit section implementing the approximation $f \cdot (1 - f)$ of the derivative of neuron activation function, to be used for evaluating δ_j (note that this would be exact if the function f were an ideal sigmoid).

3.2 Synapses

Different circuit implementations of the various multipliers present in the in the network are possible [Mea89, Vit90]. In the case of Fig. 6a, the output current I_{out} of the two-quadrant multiplier is proportional to the product of the current I_b in the weight transistor times the difference (V_{in} - V_{ref}) applied to the differential pair.

217

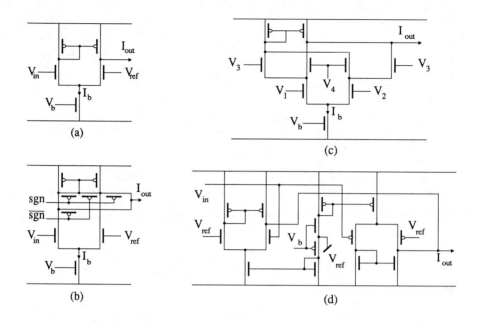

Fig. 2. Four different multipliers to be employed in the various blocks of the architecture.: (a) exponential two quadrant multiplier; (b) the same as (a) but with a digital sign input; (c) Gilbert linear multiplier; (d) four quadrant exponential multiplier.

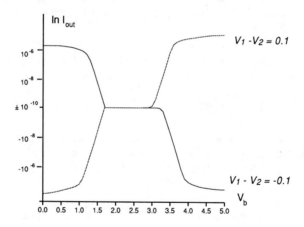

Fig. 7. - Transfer curve for the circuit of Fig. 6d.

This occurs also in the circuit of Fig. 6b, which, in addition, has a digital input (sgn) to invert the direction of the output current (if adopted for a synaptic weight multiplier this allow to change the sign of the weight).

Due to power dissipation considerations, it is useful that the synapses operate in the subthreshold region of the weight transistor. The main consequence of this fact is the exponential increase of the weight with respect to V_b. Moreover, this transistor features a width in the 100μm range to generate a minimum reasonable current of at least few nA.

If a continuous variation of the weight is necessary from negative to positive values of the weight, the usual Gilbert multiplier (Fig. 6c) can be adopted, taking into account both the linear dependence of the weight from the V_b voltage, and the almost constant power dissipation in the various operating conditions, and in particular when $W_{ij} = 0$. On the contrary, the exponential dependence of weight is maintained in the circuit of Fig. 6d. It consistes in a pair of OTA amplifiers biased alternatively in dependence of the voltage Vb. A characteristic of this circuit is shown in Fig. 7.

3.3 Multilevel storage cell

If the network operates only in the feed forward mode (for example, after the learnig process has completed), the charge on the gate of the weight transistor must be kept as much stable as possible. With this aim, to overcome the problem of leakage current, it is necessary to operate in a refresh mode, thus achieving a stable memorization of the voltage, V_b, controlling the synapses [TsSa87, IbZa90]. Since no purely analog refresh tecnique is assessed, (also if some proposals have been recently made [Ho/90]), it is necessary to introduce some sort of discretization. We propose to adopt a Multilevel Cell Storage technique as reported in [Ao/87, Ho88].

The Multi Level (ML) memory cell is the same adopted in standard DRAMS: it needs special interface circuitry for operating internal multi level voltage values (see

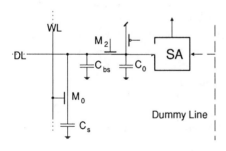

Fig. 8 (a). Basic memory circuit

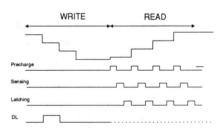

Fig. 8 (b). A four level operation scheme for the multi-level cell memory.

figure 8a). Figure 8b reports a four level operation scheme. The storage capacitor Cs is accessed by means of the transistor M_0: in practise it will be constituted by the gate capacitance of the weight transistor. The addressing signal WL (Word Line) is common to all cells of the same column. The cells of a row are connected to the same Data Line (DL): C_{bs} accounts for the parasistics capacitances of the DL. The transistor M_2 constitutes a charge preamplifier [He/76] which makes the Sense Amplifier (SA) input signal almost indipendent from the value of C_{bs}. On the other side of the SA, a dummy line gives a reference signal.

The WRITE operation is performed applying a descending staircase voltage V_p to the WL of the selected column: to store the ith-level, a positive pulse is applied to the data line at the ith step. The stored voltage V_s is equal to $V_p - V_{TM0}$, where V_{TM0} is the threshold voltage of transistor M_0.

During the READ operation, an ascending staircase is applied to the WL of the selected column. At each step of the staircase, three different phases occur: 1)precharge: the DL is set at the high value by the transistor M_1; 2) sensing: the selected column is addressed and, if the stored voltage value is lower than the level staircase, a charge transfer occurs among C_s, C_{bs} and C_0, generating a voltage drop in input to the SA; 3) latching: the SA is activated; the SA generates a signal if its input signal is large enough. This output strobes a register containing a code of the step count at which the charge transfer occurred, performing in such a way a sort of analog to digital conversion.

The REFRESH operation is performed by a READ-WRITE sequence.

Circuit simulations indicate that it is possible to store voltage levels on a capacitors of 0.1pF stepped by a few tenth millivolts. The cicle time (a READ operation followed by a WRITE operation) is in the range of 10÷20 msec.

This technique allows the memorization of a voltage ranging from about 0V to 3÷3.5V : since this is not simmetrical with respect to 2.5 V, it is necessary to adopt a dual ramp tecnique, using simmetrical circuitry. The n-section described so far is responsible for the voltages between 0 and 2.5V; the p-section, that can be derived complementing the previous one, operates between 2.5 and 5V.

In our case, the capacitance of the gate of the weight transistor is in the range of few hundreds fF, and the voltage step, that can be reasonably reached, is consequently in the range of 20÷40 mV. Since V_b varies in a range of about 0.5 below threshold, not to operate with currents under 1 nA, the total number of levels available for Vb goes from about 12 to 25, which fits the considerations made in Section 2.1.

When the network operates in the learning mode, a charge transfer tecnique such as proposed in [IbZa90] can be used to modify the value of voltage V_b in the weight transistor.

The circuits have been simulated with a version of PSPICE with an accurate model for the sub-threshold region of conduction derived from [An/82]. The delay between a voltage step at a synapse input and the corresponding neuron output has been evaluated in about 1μsec for a 65 input neuron.

4. Conclusions

A quasi-analog adaptive architecture for the Back Propagation algorithm has been described in this chapter. It has been demonstrated, through simulations, how the BP algorithm can be operated successfully with discretized weights. In particular, better performances can be achieved with an exponential discretization, i.e. the strength of weights varies exponentially with the controlling variable (voltage).

The discretized voltage values differ by a quantity high enough that the neural network can be backed up with a refresh technique in combination with a Multi-level Dynamic Memory, that entails a particularly low wiring cost.

A CMOS circuit implementation has been discussed which properly matches the modified BP algorithm. The mechanism controlling weight changes is simple enough to be reproduced locally at each synapse site thus complying with one of the properties needed by an efficient storage technology for analog VLSI adaptive systems [Sc/89].

5. Acknowledgements

The authors wish to acknowledge helpful discussions with Prof. E. Pasero of University of Rome II, Prof. R. Castello and Prof. F. Montecchi of University of Pavia and with our colleague Prof. S. Ridella.

6. References

[An/82]   P. Antognetti, D.D. Caviglia and E. Profumo, "CAD Model for Threshold and Subthreshold Conduction in MOSFET's", IEEE Jour. Solid State Circ., Vol. SC-17, No. 3, June 1982, pp. 454-458.

[AnRo88] J.A. Anderson, and E. Rosenfeld, Eds., "Neurocomputing: Foundations of Research", The MIT Press, Cambridge, MA, 1988.

[Ao/87]   M. Aoki, Y. Nakagome, M. Horiguchi, S. Ikenaga, and K. Shimishigashi, "A 16-Level/Cell Dynamic Memory", IEEE Jour. Solid State Circ., Vol. SC-22, No. 2, April 1987, pp. 297-299.

[BiIs90]   S. Bibyk and M. Ismail, "Neural Network Building Blocks for Analog MOS VLSI" in Analog IC Design: the Current-Mode Approach, C. Toumazou et al. Eds., Peregrinos Ltd., 1990.

[Bo/89]   K. Boonyanit, et al., "Analog Storage for Neural Networks Via EPROM Technology", Technical report, Stanford University, 1989.

[Br/89]   M.L. Brady et al., "Back Propagation Fails to Separate where Perceptrons Succeed", IEEE Trans. on Circ. and Syst., Vol. 36, no. 5, May 1989, pp. 665-674.

[Ca/88]   D.D. Caviglia, G.M. Bisio and G. Parodi, "Neural Algorithms on VLSI Concurrent Architectures", Int. Neural Network Conf. INNS 88, Boston, Sep. 1988, p. 377.

[Ca/90]   D.D. Caviglia, M. Valle and G.M. Bisio, "Effects of Weight Discretization on the Back Propagation Learning Method: Algorithm Design and Hardware Realization", Proc. International Joint Conference on Neural Networks, IJCNN 90 San Diego, June 1990.

[He/76]   L.G. Heller, et al., "High Sensitivity Charge-Transfer Sense Amplifier",Journal of Solid State Circuits, Vol. SC-11, No. 5, October 1976.

[Ho/88]   M. Horiguchi, M. Aoki, Y. Nakagome, S. Ikenaga, and K. Shimishigashi, "An Experimental Large-Capacity Semiconductor File Memory Using 16-Levels/Cell Storage", IEEE Jour. Solid State Circ., Vol. SC-23, No. 1, Feb. 1988, pp. 27-33.

[Ho/90]   Y. Horio, M. Ymamamoto and S. Nakamura,"Active Analog Memories for Neuro-Computing", IEEE International Symposium on Circuits and Systems, ISCAS 90, New Orleans, May 1990, pp. 2988-2989.

[Hoc89]   B. Hochet, "Multivalued MOS Memory for Variable Synapse Neural Networks", Electronics Letters, Vol. 25, No. 10, 11 May 1989.

[IbZa90]   F. Ibrahim and E.M. Zaghloul, "Design of Modifiable-Weight Synapse CMOS Analog Cell", IEEE International Symposium on Circuits and Systems, ISCAS 90, New Orleans, May 1990, pp. 2978-2981.

[IEMI89]   IEEE Micro, Special issue on "Silicon Neural Networks", Dec. 1989.

[KoPo84]   C. Koch and T. Poggio, "Biophysics of Computation: Neurons Synapses and Membranes", A.I. Memo 795 MIT, October 1984.

[Mea89]   C.A. Mead, "Analog VLSI and Neural Systems", Addison Wesley, 1989.

[MeIs89]   C. Mead and M. Ismail, Eds. , "Analog VLSI Implementation of Neural Systems", Kluwer Academic Publisher, 1989.

[MeMa88]   C.A. Mead and M.A. Mahowald, "A Silicon Model of early Visual Processing", Neural Networks, Vol. 1, No. 1, 1988.

[McPi43]   W.S. Mc Culloch, and W.Pitts, "A Logical Calculus of Ideas Immanent in Neural Nets", Bull. of Math. Biophysics, Vol. 5, 1943, pp. 115-133.

[PiMe47]   W. Pitts, and W.S. Mc Culloch, "How We Know Universals: the Perception of Auditory and Visual Form", Bull. of Math. Biophysics, Vol. 9, 1947, pp. 127-147.

[RuMc86]   D.E. Rumelhart and J.L. McClelland (Eds.), "Parallel Distributed Processing", MIT Press, Cambridge, Mass., 1986.

[Sc/89]   D.B. Schwartz, R.E. Howard and W.E. Hubbard, "A Programmable Analog Neural Network Chip", IEEE J. Solid-State Circuits,Vol. 24,No. 2, pp. 313-319, 1989.

[TzSa87]   Y. Tzividis and S. Satyanarayana, "Analogue Circuits for Variable-Synapse Electronic Neural Networks", Electronics Letters, 19th November 1987, Vol. 23, No. 24 pp. 1313-1314.

[Vit85]   E.A. Vittoz, "Micropower Techniques", in Design of MOS VLSI Circuits for Telecommunications, Y. Tzividis and P. Antognetti, Eds., Prentice Hall 1985.

[Vit90]   E.A. Vittoz, "Analog VLSI Implementation of Neural Networks", IEEE International Symposium on Circuits and Systems, ISCAS 90, New Orleans, May 1990, pp. 2524-2527.

[VNeu58]   J. Von Neumann, " The Computer and the Brain", Yale University Press, 1958.

[Vo/88]   T.P. Vogl et al., "Accelerating the Convergence of the Back-Propagation Method", Biol. Cybern., Vol. 59, pp.257-263, 1988.

9

THE TINY-TANH NETWORK: TOWARDS CONTINUOUS-TIME SEGMENTATION WITH ANALOG HARDWARE

John G. Harris

Computation and Neural Systems Program, 216-76
California Institute of Technology
Pasadena, CA, 91125

and

Hughes Artificial Intelligence Center
3011 Malibu Canyon Rd.
Malibu, CA 90265

Segmentation is a basic problem in computer vision. We introduce the tiny-tanh network, a continuous-time network that segments scenes based upon intensity, motion, or depth. The tiny-tanh algorithm maps naturally to analog circuitry since it was inspired by previous experiments with analog VLSI segmentation hardware. A convex Lyapunov energy is utilized so that the system does not get stuck in local minima. No annealing algorithms of any kind are necessary—a sharp contrast to previous software/hardware solutions of this problem.

1 Introduction

A large class of vision algorithms is based on minimizing an associated cost functional. Such a variational formalism is attractive because it allows a priori constraints to be stated explicitly. The single most important constraint is that the physical processes underlying image formation—such as depth, orientation, and surface reflectance—change slowly in space. For instance, the depths of neighboring points on a surface are usually similar. Standard regularization algorithms embody this smoothness constraint and lead to quadratic variational functionals with a unique, global minimum (Horn and Schunck, 1981; Hildreth, 1984; Poggio, Torre and Koch, 1985; Poggio, Voorhees and Yuille, 1986; Grimson, 1981). These quadratic functionals can be mapped onto linear resistive networks, such that the stationary voltage distribution, corresponding to the state of least power dissipation, is equivalent to the solution of the variational functional (Horn, 1974; Poggio and Koch, 1985). Inputs are supplied by injecting currents into the appropriate nodes.

Smoothness breaks down, however, at discontinuities caused by occlusions or by differences in the physical processes underlying image formation (such as different surface reflectance properties). Detecting these discontinuities becomes crucial, not only because otherwise smoothness is applied incorrectly but also because the locations of discontinuities are often required for further image analysis and understanding (for example, we can often find the outline of a moving object reliably by detecting discontinuities in the optical flow field). Geman and Geman (1984) first introduced a class of stochastic algorithms, based on Markov random fields, that explicitly encodes the absence or presence of discontinuities by means of binary variables. Their approach was extended and modified by numerous researchers to account for discontinuities in depth, texture, optical flow, and color. An appropriate energy functional is minimized using stochastic optimization techniques, such as simulated annealing. Various deterministic approximations, based on continuation methods or on mean field theory yield next-to-optimal solutions (Koch, Marroquin and Yuille, 1986; Terzopoulos, 1986; Blake and Zisserman, 1987; Hutchinson et al., 1988; Blake, 1989; Geiger and Girosi, 1989).

In the 1-D case the sparse and noisy depth data d_i are given on a discrete grid. Associated with each lattice point is the value of the recovered surface u_i and a binary line discontinuity ℓ_i. When the surface is expected to be smooth except at isolated discontinuities, the functional to be minimized is given by

$$J(u, \ell) = \sum \left[(d_i - u_i)^2 + \lambda^2 (u_{i+1} - u_i)^2 (1 - \ell_i) + \alpha \ell_i \right] \tag{1}$$

where λ and α are free parameters. The first term, with the sum including only those locations i where data exist, forces the surface u to be close to the measured data d. The second term implements the piecewise smooth constraint: if all variables, with the exception of u_i, u_{i+1}, and ℓ_i, are held fixed and $\lambda^2(u_{i+1} - u_i)^2 < \alpha$, then it is "cheaper" to pay the price $\lambda^2(u_{i+1} - u_i)^2$ and to set $\ell_i = 0$ than to pay the larger price α; if the gradient becomes too steep, $\ell_i = 1$, and the surface is segmented at that location. The final surface u is the one that best satisfies the conflicting demands of piecewise smoothness and fidelity to the measured data.

2   Resistive Fuse

Analog circuits provide an interesting and powerful computational medium for developing early vision algorithms. As these small, inexpensive, low-power chips become common, researchers will possess real-time prototyping capabilities to address fundamental issues. Mead (1989) has pioneered the use of subthreshold analog CMOS for early sensory processing. For a review of analog circuits applied to early vision, see Horn (1989) and Koch (1989).

Instead of resorting to stochastic search techniques to find the global minimum of Eq. 1, we use a deterministic approximation and map the functional J onto the circuit shown in Fig. 1. The stationary voltage at every gridpoint then corresponds to u_i. If data exist at location i, a battery is set to d_i. The conductance between the battery and the grid is assumed to be 1 where data exist and 0 if no data exist at a particular location i. In the absence of any discontinuities (all $\ell = 0$), smoothness is implemented via a conductance of value λ^2 connecting neighboring grid points; that is, the nonlinear resistors in Fig. 1 can simply be considered linear resistors. The cost functional J can then be interpreted as the power dissipated by the circuit. If parasitic capacitances are added to the circuit, J acts as a Lyapunov function of the system and the stationary voltage distribution corresponds to the smooth surface.

Previously, we designed a two-terminal nonlinear device, which we call a resistive fuse, to implement piecewise smoothness (Harris *et al.*, 1990a). If the magnitude of the voltage drop across the device is less than $V_T = (\alpha^{1/2}/\lambda)$, the current through the device is proportional to the voltage, with a conductance of λ^2. This implements smoothness. If V_T is exceeded, the fuse breaks and the current goes to zero. Unlike the common electrical fuses in houses, the operation of the resistive fuse is fully reversible. We use an analog fuse with the I-V curve shown in Fig. 2, implementing a continuous version of the binary line discontinuities[1]. If the internal dynamics of the resistive fuse can be neglected, then

---

[1] Perona and Malik (1989) discuss the performance of similar devices using anisotropic diffusion.

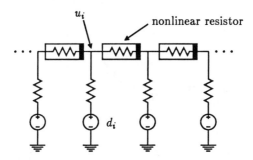

Figure 1: Schematic diagram for our hardware segmentation networks. A mesh of non-linear resistive elements (shown as resistors inside rectangles) provides the smoothing and segmentation ability of the network. The data are given as battery values d_i with a unit conductance connecting each battery to the grid. In the absence of any discontinuities, the nonlinear resistors have a conductance of λ^2. The output is the voltage u_i at each node.

it can be proven that our network will not oscillate, but rather will settle into a local minimum. The associated Lyapunov function is the electrical co-content (Harris *et al.*, 1989).

We built a 20 by 20 pixel VLSI chip, using the subcircuit types and design practices developed by Mead (1989). The slope λ^2 and the voltage threshold of all fuses can be set by off-chip voltage inputs. Figure 3a shows a figure eight pattern that was scanned into the chip. The height of the signal was 0.5V with evenly-distributed additive noise of ±0.25V. Figure 3b shows the measured voltage values that were scanned off the chip. These voltages could represent image intensity, depth, or motion. Algorithm simulations have shown that the resistive fuse idea can be combined successfully with a motion algorithm to compute segmented optical flow fields from digitized images (Harris *et al.*, 1990b).

Mapping nonlinear computations to analog VLSI does not avoid the problem of local minima. Several continuation methods have been developed that gradually warp the Lyapunov energy of the system from a convex one to the final desired energy (Harris *et al.*, 1989, Lumsdaine *et al.*, 1990). Continuation methods rely on the fact that the unique minima of the convex energy are close to the global minima of the desired energy. Unfortunately, these methods require clocking circuitry to vary an on-chip voltage through a range of values. Furthermore, since our analog technology naturally implements continuous-time systems with on-chip photoreceptors, it would be awkward to resort to a sampled-time system. This problem is especially evident when we try to recover the op-

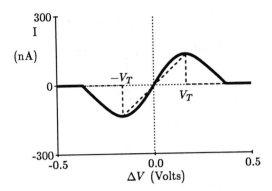

Figure 2: The solid line shows the measured current-voltage relationship for the resistive fuse. The I-V curve of a binary fuse is also illustrated. For a voltage of less than V_T across this two-terminal device, the circuit acts as a resistor with conductance λ^2. Above V_T, the current is either abruptly set to zero (binary fuse) or smoothly goes to zero (analog fuse). Independent voltage control lines allow us to change both λ^2 (over four orders of magnitude) and V_T (from 50 to 500 mV). In addition, we can vary the I-V curve continuously from the hyperbolic tangent of Mead's saturating resistor (HRES) to that of an analog fuse.

tical flow in the presence of motion discontinuities (Harris *et al.*, 1990b). Motion already contains a natural time scale that we would have to ensure is much longer than the time scale of our continuation methods.

3   Tiny Tanh Network

Rather than deal with resistive networks that have many possible stable states, we propose to utilize a network that has a single unique stable state. Any circuit made of independent voltage sources and two-terminal resistors with strictly increasing I-V characteristics has a single unique stable state. This result has been well known; a proof is given in Chua *et al.* (1987). Gradient descent algorithms are guaranteed to find this solution. Figure 4 shows a plot of our proposed element's I-V characteristic, a simple hyperbolic tangent relationship with an adjustable width and slope. The slope in the linear region is defined to be λ^2, and the curve saturates for voltage drops greater than δ. Since the derivative of the tanh function is always positive, the nonlinear resistive network containing them is guaranteed to converge to a unique minimum. A piecewise linear approximation of this curve was used in simulations and analysis. This saturating element is used as a the

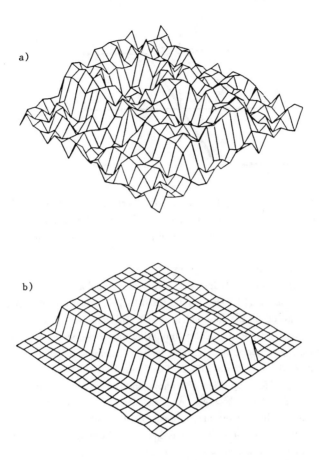

Figure 3: Experimental data from the two-dimensional resistive fuse chip. (**a**) A noisy figure eight pattern was used as input. The height of the signal is 0.5V with evenly distributed additive noise of ±0.25V. (**b**) The measured voltage output scanned off the chip.

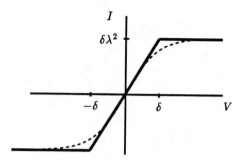

Figure 4: The desired saturating I-V characteristic is shown as a dotted line. Its piecewise linear approximation (solid line) was used for simulation and analysis. The slope of the linear region (λ^2) and the saturation threshold (δ) are adjustable parameters.

nonlinear resistive element shown in Figure 1.

Mead has constructed a saturating resistor in subthreshold analog VLSI (Mead, 1989). This device is often called HRES (Horizontal RESistor) since it was originally designed to model the horizontal cells in the mammalian retina. To evaluate how the saturating resistor performs, a fourth-order Runge-Kutta simulation was developed for a two-dimensional network of saturating resistors. A unit step edge was generated for a 20 x 20 pixel image and gaussian noise with $\sigma = .2$ was added. The filled circles in Figure 5 represent a 1-D cross section through this array. Figure 5a shows the typical segmentation result for the saturating resistor, $\delta = .1$ and $\lambda^2 = 4.0$. As Mead has observed, a network of saturating resistors performs an edge enhancement (Mead, 1989). Unfortunately, the noise is still evident in the output, and the curves on either side of the step have started to slope towards one another. As λ is increased to smooth out the noise, the two sides of the step will blend together into one homogeneous region. These same results are observed with hardware resistive networks using Mead's saturating resistor. These saturating resistors cannot segment data into regions of roughly uniform voltage.

The new idea proposed here is that the saturating resistor can be run in a very different region of operation where network segmentation properties are greatly enhanced. If we decrease the width of the linear region significantly ($\delta = .005$) while simultaneously increasing the conductance ($\lambda = 16$), the result shown in Fig. 5b is produced. The height of the step[2] detected is 200 times larger than δ. Figure 6 shows a simulation of the tiny-

---

[2]If the signal size was 100mV, for example, a resistor with a linear region with a total width of 1mV would be required to provide the same performance. Mead's saturating resistor has a minimum linear region width of about 100mV.

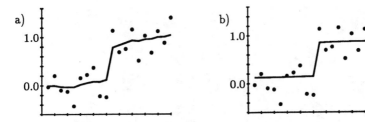

Figure 5: A 20x20 pixel unit step edge was generated, and gaussian noise with $\sigma = .2$ was added. The filled-circles indicate a 1-D cross section through this input array. **(a)** Shows the segmentation result for the saturating resistor in the usual setting, $\delta = .1$ and $\lambda = 2.0$. For these parameter settings, the noise is still visible in the output and the curves on either side of the step have started to slope towards one another. **(b)** The enhanced region of operation ($\delta = .005$ and $\lambda^2 = 256$) segments the input into two distinct uniform regions.

tanh network segmenting the famous mandrill image for various values of δ. The network segments the image into blocks of approximately uniform brightness with δ determining the scale of the computation. No continuation method or annealing strategy of any sort was necessary to produce these simulation results.

We can understand why the network performs so well by studying the step response of a 1D network. Figure 7a shows the input of a step to the network of step size h. The nodes labeled V_L and V_H are the nodes to the immediate left and right of the step. If we use the piecewise linear approximation to the tiny-tanh, there are two cases to consider for the ideal step input. The resistive element between nodes V_L and V_H is either in its linear or saturation region of operation. For small input steps, the network is effectively linear. For $\lambda \gg 1$, we find that $V_L \approx V_H \approx h/2$. For larger inputs, one resistive element can be replaced by a constant current source of value $\delta\lambda^2$. For $\lambda \gg 1$, we find that $V_L \approx \delta\lambda$ and $V_H \approx h - \delta\lambda$. By setting the voltages in the two regimes equal we find that the breakpoint occurs at $h \approx 2\delta\lambda$. Figure 7b depicts the step response of the tiny-tanh network.

Notice that the network does not recover the exact heights of input edges. Rather it subtracts a constant ($2\delta\lambda$) from the height of each input. This effect is noticeable in Fig. 5b where the unit step input has decreased slightly in magnitude. In two dimensions, this effect becomes more interesting. For regions with small area to circumference ratios, the step height will decrease by a larger amount. Typically, the exact values of the heights are less important than the location of the discontinuities. Furthermore, it would not be difficult to construct a two-stage network to recover the exact values of the step heights if

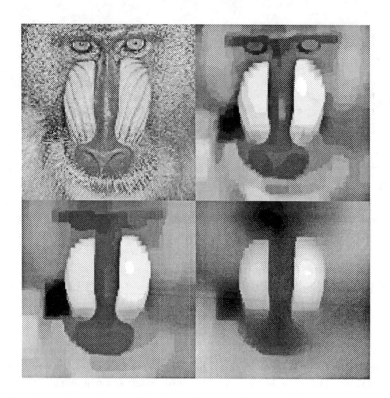

Figure 6: Simulations of tiny-tanh network on the mandrill image. $\lambda = 32$ for all four images. Top left is the original ($\delta = 0$), top right $\delta = 0.0064$, bottom left $\delta = 0.0128$ and bottom right $\delta = 0.0512$. The original image consisted of 256x256 8-bit pixels. Each pixel ranged in value from 0 to 255. The network segmented the image into blocks of approximately uniform brightnes with δ determining the scale of the computation.

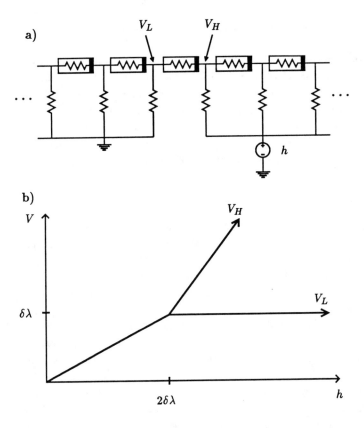

Figure 7: a) Schematic of step input of height h for the nonlinear network shown in Fig. 1. **b)** Plot of voltage V_H and V_L versus h assuming $\lambda \gg 1$. Inputs less than $2\delta\lambda$ are smoothed out since $V_H - V_L \approx 0$. Inputs steps larger than $2\delta\lambda$ are recovered as step edges with $V_H - V_L \approx h - 2\delta\lambda$.

desired. In this scheme a tiny-tanh network would control the switches on a second fuse network.

In retrospect, we have changed the nonlinear line-process computation to a simpler problem that can be solved by minimizing a convex energy. Notice that the network does not have an explicit representation of discontinuities. It merely subtracts a fixed voltage $(2\delta\lambda)$ from the height of input step edges. If a binary discontinuity map is desired, a threshold must be chosen for an additional post-processing step. Unlike typical line-process energy minimization algorithms, the whole segmentation computation is accomplished without defining a binary threshold for edges.

4   Conclusion

A rigid piecewise-constant assumption about the world is unnecessarily restrictive. The continuous-time network simulated here could be used to implement a thin plate energy and segment based upon higher order discontinuities (Grimson, 1981; Terzopoulos, 1983; Blake and Zisserman, 1987; Harris, 1989; Liu and Harris, 1989). A successful extension of these methods will allow depth and intensity to be segmented into piecewise linear regions.

We have discussed a hardware resistive fuse network that segments inputs in real-time. We proposed a much simpler tiny-tanh network that performs similar computations in continuous-time. The paper illustrates two of the fundamental advantages of using analog circuits to experiment with vision algorithms. This new computational medium gives us the ability to deal with time explicitly and rapidly prototype nonlinear systems. The resistive fuse network is a highly nonlinear system which has forced us to think explicitly about dealing with time in our solution strategies. Continuation methods are awkward to implement in continuous-time fashion and so alternative methods are desirable. Through experimentation with analog hardware, we have discovered an algorithm that is useful for any sort of implementation—analog or digital.

Acknowledgments

This work was done in close cooperation with Christof Koch. Thanks to Carver Mead for making this research possible. All chips were fabricated through MOSIS with DARPA's support. This research was partially supported by NSF grant IST-8700064, a grant from the Office of Naval Research, by DDF-II funds from the Jet Propulsion Laboratory at the California Institute of Technology.

5 References

Blake, A. (1989). Comparison of the efficiency of deterministic and stochastic algorithms for visual reconstruction. *IEEE Trans. Pattern Anal. Mach. Intell.* **11**:2–12.

Blake, A. and Zisserman, A. (1987). *Visual Reconstruction.* Cambridge, MA: MIT Press.

Chua, L. O., Desoer, C. A., and Kuh, E. S. (1987). *Linear and Nonlinear Circuits.* New York: McGraw-Hill, pp. 23–34.

Geiger, D. and Girosi, F. (1989). Parallel and deterministic algorithms from MRF's: surface reconstruction and integration, MIT AI Memo 1114, May 1989.

Geman, S. and Geman, D. (1984). Stochastic relaxation, Gibbs distribution and the Bayesian restoration of images. *IEEE Trans. Pattern Anal. Mach. Intell.* **6**:721–741.

Grimson, W. E. L. (1981). *From Images to Surfaces.* Cambridge, MA: MIT Press.

Harris, J. G. (1989). An analog VLSI chip for thin plate surface interpolation. In *Neural Information Processing Systems I*, ed. D. Touretzky. Palo Alto: Morgan Kaufmann.

Harris, J. G., Koch, C., Luo, J. and Wyatt, J. (1989). Resistive fuses: analog hardware for detecting discontinuities in early vision. In: *Analog VLSI Implementations of Neural Systems*, Mead, C. and Ismail, M. eds., Kluwer, Norwell, MA.

Harris, J. G., Koch, C., and Luo, J. (1990a). A two-dimensional analog VLSI circuit for detecting discontinuities in early vision. *Science* **248**:1209-11.

Harris, J. G., Koch, C., Staats, E., and Luo, J. (1990b). Analog hardware for detecting discontinuities in early vision. *Internat. Journal of Comp. Vision* 4:211-223.

Hildreth, E.C. (1984), *The Measurement of Visual Motion.* MIT Press, Cambridge, MA.

Horn, B.K.P. (1974). Determining lightness from an image. *Compt. Graph. Imag. Processing* **3**:277-299 (1974).

Horn, B. K. P. and Schunck, B. G. (1981). Determining optical flow. *Artif. Intell.* **17**:185–203.

Horn, B. K. P. (1989). Parallel networks for machine vision. *Artif. Intell. Lab. Memo No.* **1071**, MIT, Cambridge, MA.

Hutchinson, J., Koch, C., Luo, J., and Mead, C. (1988). Computing motion using analog and binary resistive networks. *IEEE Computer* **21**:52–63.

Koch, C., Marroquin, J., and Yuille, A. (1986). Analog "neuronal" networks in early vision. *Proc. Natl. Acad. Sci. USA* **83**:4263–4267.

Koch, C. (1989). Seeing chips: analog VLSI circuits for computer vision. *Neural Computation* **1**:184–200.

Liu, S. C. and Harris, J. G. (1989). Generalized smoothing networks in solving early vision problems. *Proc. IEEE Computer Vision and Pattern Recognition Conference*, San Diego, CA, June 1989, pp. 184-191.

Lumsdaine, A., Wyatt, J. and Elfadel, I. (1990), Nonlinear Analog Networks for Image Smoothing and Segmentation, Proc. 1990 *IEEE Internat. Symp. on Circuita and Systems*, New Orleans, LA, May 1990, pp. 987-991.

Mead, C. A. (1989). *Analog VLSI and Neural Systems*. Reading: Addison-Wesley.

Perona, P. and Malik, J. (1988). A network for multiscale image segmentation. *Proc. 1988 IEEE Int. Symp. on Circuits and Systems*, Espoo, Finland, June, pp. 2565–2568.

Poggio, T. and Koch, C. (1985). Ill-posed problems in early vision: from computational theory to analogue networks. *Proc. R. Soc. Lond. B* **226**:303–323.

Poggio, T., Torre, V., and Koch, C. (1985). Computational vision and regularization theory. *Nature* **317**:314–319.

Poggio, T., Voorhees, H., and Yuille, A. (1986). A regularized solution to edge detection. Artif. Intell. Lab Memo **No. 833** (MIT, Cambridge).

Terzopoulos, D. (1983). Multilevel computational processes for visual surface reconstruction. *Comp. Vision Graph. Image Proc.* **24**:52–96.

Terzopoulos, D. (1986). Regularization of inverse problems involving discontinuities. *IEEE Trans. Pattern Anal. Machine Intell.* **8**:413–424.

10

Digital Neural
Network Implementations

James B. Burr

Department of Electrical Engineering
Stanford University
Stanford, Ca. 94305

Abstract

This chapter gives an overview of existing digital VLSI implementations and then discusses techniques for implementing high performance, high capacity digital neural nets. It presents a set of techniques for estimating chip area, performance, and power consumption in the early stages of design to facilitate architectural exploration. It shows how technology scaling rules can be included in the estimation process. It presents a set of basic building blocks useful in implementing digital networks. It then uses the estimation techniques to predict capacity and performance of a variety of digital architectures. Finally, it discusses implementation strategies for very large networks.

1 Introduction

Neural network applications suitable for implementation in VLSI cover a wide spectrum, from dedicated feedforward nets for real time feature detection to general purpose engines for exploring learning algorithms. The DARPA Neural Network Study [76] contains a good discussion of the range of applications together with a method of classifying them which we use in this chapter.

The fastest analog neurochips are almost six orders of magnitude faster than an average workstation or PC, and three orders of magnitude faster than the fastest supercomputer. The fastest digital neurochip is about as fast as a supercomputer. Even so, many of the potential applications demand far more computational power and synaptic storage capacity than has been implemented so far. As progress is made toward implementation of larger, higher performance networks, and as technology scales further into the submicron regime, power dissipation will become a dominant constraint, and effective power analysis and optimization will be an important aspect of design.

There is always a tradeoff in VLSI implementation between performance and flexibility. Both are needed in neural nets. Mature applications with well characterized algorithms will continue to seek higher performance as learning algorithm research continues to suggest new architectures.

This chapter discusses both existing implementations and some of the challenges facing future implementations. Section 2 proposes an informal classification scheme for neurochips. Section 3 describes some of the more recent digital VLSI implementations. A table at the end of section 3 gives a quick overview of implementations to date. Section 4 discusses architectural tools which can be leveraged in implementing high performance, high capacity networks. Section 5 describes some low level basic building blocks useful in digital implementations. Section 6 describes techniques for estimating area, performance, and power of a desired architecture, and shows how technology scaling rules can be included in the estimation process. Section 7 discusses techniques for maximizing performance on a constrained power budget. Section 8 discusses the impact of technology scaling on future neurochip architectures. Section 9 discusses a chip we have been working on at Stanford. Section 10 discusses obstacles and opportunities to implementing large digital networks.

2 Classifying VLSI implementations

Yutaka Akiyama's PhD thesis [4] has a comprehensive section on neurochip classification. Since its publication, the number and variety of neurochips has proliferated. We can group these chips into major categories depending on whether they use analog or digital computation, how synaptic storage is implemented, whether they have some form of on-chip learning, and whether they are standalone or are meant to be connected to other chips to implement larger networks.

The most common analog architecture is a fully connected mesh of N^2 synaptic processors supporting N neurons. This architecture can perform feedforward computations at very high rates, typically converging in a few microseconds, and is well suited to Hopfield nets [38] and associative memories [25, 24]. For example, ATT's associative memory chip [24] has 256 neurons and 65536 synapses. Assuming 10 iterations through the network for convergence, this chip has a feedforward computation rate of around 500 GCPS (billion connections per second).

Another analog architecture used in the Caltech retina chip [71] is a lattice of neural processors with nearest neighbor connections. This architecture can also compute at very high rates. The chip has 2304 neurons, each connected to 6 neighbors. Assuming convergence in a few microseconds, this chip has a feedforward computation rate of around 10 GCPS.

Digital chips tend to be organized as a 1 dimensional systolic array [47, 46, 45]. They generally have lower feedforward computation rates than analog chips; values reported range from 3 MCPS [21] to 1.28 GCPS [27].

Analog implementations usually compute the inner product as an analog current sum, but differ in how the weights are stored. Techniques reported include resistors [24, 25, 71], CCDs

[2, 14, 65], capacitors [18, 48, 56], and floating gate EEPROMs [37].

Digital implementations use either parallel, bit-serial, or stochastic arithmetic. Some have one processor for each synapse; most have one processor per neuron or even one processor for many neurons. Synaptic weights are either stored in shift registers, latches, or memory. Memory storage alternatives include one-, two-, or three-transistor dynamic RAM, or four- or six-transistor static RAM.

Most digital implementations so far include some form of on-chip learning; most analog implementations do not.

A few hybrid chips have been reported which combine analog and digital techniques [7, 55, 23]. One common approach is to store weights digitally and to compute the inner product as a current sum.

Most chips that learn store weights digitally. One notable exception is a purely analog Boltzmann chip [9], which uses its learning algorithm to refresh capacitive weights. This chip is implemented in 1.0μ 2-poly 2-metal CMOS. It has 125 neurons and 10000 capacitive synapses. Each synapse implements the Boltzmann weight update rule [33, 1] with 42 transistors, and measures $200\lambda \times 200\lambda$. Network convergence in response to a single input takes 5 microseconds, including annealing, which is implemented as a damped oscillation on a reference voltage. Assuming a 50 step anneal schedule, the equivalent digital feedforward computation rate would be about 100 GCPS.

3 Existing digital implementations

3.1 TI's NETSIM

NETSIM [21] was one of the first purely digital architectures reported. It consists of two types of chips: a "solution engine" chip and a "communication handler" chip. Neural computation is done in the solution engine; neural activations are routed through the communication handler. The solution engine requires external memory to store weight values. A solution engine, a communication handler, a local microprocessor, and memory are packaged on a single NETSIM card. A system consists of a number of NETSIM cards in a 3-D array interfaced to a general purpose computer.

The solution engine performs an 8x8 bit multiply-accumulate in 250ns, so the performance is 4 MCPS.

These chips do not appear to have been implemented, but the architectural ideas appear in various forms in later chips. The idea of pipelined off chip weights is especially appealing for very large networks. Also, considerable attention was given in the design to multichip implementations.

3.2 Duranton and Sirat's digital neurochip

Duranton and Sirat's proposed chip [17] is a fully digital architecture which stores 16 bit synaptic weights in an on-chip RAM. The architecture supports on-chip learning, but exports the sigmoid function. It uses an interesting bit-serial technique to form the inner product.

On each cycle, one bit of each activation is anded with the weights to form a set of partial products which are reduced to a single partial product in a tree of adders. Activations are 8 bits, so it takes 8 cycles to form $x_i = \Sigma w_{ij}x_j$ for a single i. The weights are fetched in parallel out of the RAM, so the RAM access time can be 1/8 the processor cycle time.

The performance estimations are given for 1.6 micron CMOS. The paper states that a new x_i is computed every 2 μsec. This implies a clock frequency of 25 MHz, and with 32 16-bit synaptic weights in parallel on every cycle, 800 MCPS. This chip also does not appear to have been implemented.

3.3 Hirai et al's digital neuro-chip

Yuzo Hirai and colleagues [35] fabricated a digital neurochip using a 1.2 micron CMOS gate array. The chip implements six neurons and 84 6-bit synapses using a variant of pulse-stream arithmetic (see section 5.7. Rather than use a chopping clock as in [57], or a comparator and pseudo-random number generator as in STONN [78], TInMANN [15], or DNNA [42], Hirai et al supply each neuron a separate, asynchronous clock to reduce pulse collisions. Each neuron requires 1 msec to evaluate its inputs. This "time constant" depends only on the number of bits in the synapses and not on the number of synapses. This implies a performance of 84 KCPS. Other pulse-stream implementations achieve better performance by allowing the network dynamics to overlap the neural computation.

3.4 Ouali and Saucier's Neuro-ASIC

Ouali and Saucier [59] have implemented a single neuron on a chip in 2.0 micron CMOS which runs at 20 MHz. The chip is a "cascadable neuron processor" that includes a local memory for storing synaptic weights and activation function parameters, a multiplier, an adder/subtractor, a controller, and input, output, and state registers for interfacing to a multichip network. The paper does not specify precision or performance, though the chip has been fabricated as a MOSIS TinyChip [73].

3.5 Neural Semiconductor's DNNA

Tomlinson et al [42] have implemented a scalable neural net architecture (DNNA: digital neural network architecture) based on stochastic pulse trains. They have implemented two chips. The SU3232 contains 1024 synaptic elements arranged in a 32×32 matrix. The NU32 chip implements 32 neurons. A fully connected network with $32N$ neurons requires N^2 synapse chips.

Each synapse includes a separate stochastic pulse train generator. Each synaptic pulse stream is anded with an activation output stream to produce a synaptic product. Synaptic products are then wire-or'ed to produce activation input streams.

The neural activation function is implemented by anding excitatory and inhibitory activation input streams. Because the wire-or operation is naturally saturating, no extra sigmoid unit is needed to implement a nonlinear activation function.

A two-chip set can implement 32 neurons and 1024 connections. k bit accuracy in the activations requires 2^k cycles. The chips run at 25 MHz, and can process 100K activations per second at 8 bit resolution. This implies a feedforward computation rate of 100 MCPS, (200MCPS if the excitatory and inhibitory parts of each synapse are counted separately).

3.6 North Carolina State's STONN

Wike et al [78] have proposed a 100K transistor CMOS Hopfield network (STONN: stochastic neural net) using stochastic logic and bit-level pipelining. They leverage the recurrent nature of Hopfield nets to overlap the neural computation with network relaxation so that by the kth iteration the neural activation has been computed to a precision of $\log_2(k)$ bits. DNNA could do this as well.

Unlike the DNNA chip, which generates stochastic samples for N^2 weights on every cycle, STONN stores weights in an on-chip shift register and generates stochastic samples of N weights per cycle. This technique achieves much higher synaptic storage density than the DNNA approach, especially if the shift register weight store is implemented as a RAM.

A weight is converted to a pulse stream by comparing it to a sequence of random numbers, using the carry out of the most significant bit of a bit-level pipelined comparator to generate the pulses. The synaptic pulse stream is "multiplied" by an input activation stream by anding the two streams. At this point, rather than wire-or the modulated input activations as in DNNA, the modulated input activation increments or decrements a counter containing the accumulating output activation. The chip has one activation comparator which generates a single stochastic sample of one of the accumulating output activations each cycle.

Weights are 7 bits; comparators are 6 bits, activation accumulators are 10 bits. The weight store is implemented using a 6-transistor dynamic shift register. The chip has 20 neurons, and 100 7 bit weights per neuron. It is projected to run between 10 and 25 MHz (it has been layed out but not fabricated). At 10 MHz, one chip can achieve 200 MCPS; at 25 MHz, 500 MCPS.

Each iteration only involves one stochastic sample of weights and activations. This type of optimization network converges over time to a stable state. STONN takes N cycles to compute the outputs of N neurons. This is the same amount of time required by a parallel implementation with one processor per neuron. The stochastic nature of the computation is much more efficient. The parallel computation is too exact. That is, one stochastic sample from one neuron can be generated each cycle. Parallel algorithms also generate one sample per cycle, but to higher precision and requiring more hardware.

Reduced precision parallel computation can require comparable hardware. The multiplier-accumulator in STONN consists of an and gate, a 6-bit comparator, a random number generator, and a 10-bit up-down counter. The multiplier in the Stanford Boltzmann chip (see section 9) is a 5-bit full adder (shared Booth encoding reduces 5 partial products to 3) and a 10-bit carry-save accumulator. The 5-bit full adder matches STONN's 6-bit comparator. The accumulator matches STONN's random number generator. The carry-save accumulator matches STONN's up-down counter.

Given bit-level pipelining, STONN should run closer to 100 MHz than 10 MHz in 2.0 micron CMOS, unless power consumption is a problem, implying 5 GCPS. It should also be able to accommodate around 64 neurons and 32K weights using one-transistor DRAM.

3.7 North Carolina State's TInMANN

TInMANN (The integer markovian artificial neural network) [15] is a stochastic architecture proposed by the STONN group to implement competitive learning using stochastic computation. The system stochastically updates the weights with a probability proportional to the neural input, causing neurons closest to an input vector to move toward it and push others away. The algorithm replaces the proportional weight update of the standard Kohonen algorithm [44] with a probabilistic update that integrates to the same value over many cycles without requiring a multiplier.

The authors predict 15 MHz operation, and 145,000 two-dimensional training examples per second.

3.8 Adaptive Solution's X1 (CNAPS)

The Adaptive Solutions X1 chip (renamed CNAPS) [27] is a general purpose SIMD multiprocessor architecture [47, 46, 45] developed for neurocomputing applications. It achieves high performance (1.6 GCPS inner product, 1.28 GCPS and 300 MCUPS doing backpropagation, 12.8 GCPS using 1-bit weights) and can implement a wide variety of learning algorithms, as well as a wide variety of signal processing algorithms. A single chip has a fairly large synaptic capacity, storing 2 million 1-bit, 256K 8-bit, or 128K 16-bit weights equally distributed among 64 processors. Multiple chips can be assembled in various topologies.

The chip (see Fig 1) contains 64 processors, a 32 bit instruction bus, an 8 bit global output bus, an 8 bit global input bus, and a 4 bit inter-processor bus. Each processor includes 4k bytes of weight memory implemented using four-transistor SRAM, a memory base address unit, 32 16-bit registers, an input unit, an output unit, an 8x16 bit multipier, a 32 bit saturating adder, and a logic and shift unit.

242

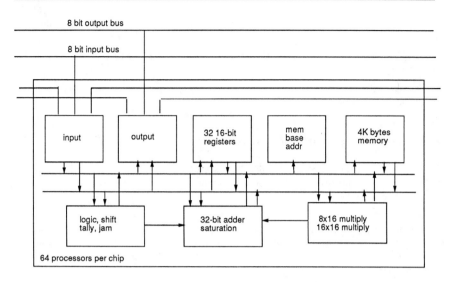

32 bit instruction bus

8 bit output bus

8 bit input bus

| input | output | 32 16-bit registers | mem base addr | 4K bytes memory |

logic, shift tally, jam | 32-bit adder saturation | 8x16 multiply 16x16 multiply

64 processors per chip

25 MHz
64 processors/chip
1.6e9 CPS
1.2e9 BP CPS
0.3e9 BP CUPS
die size: several cm^2

0.8u CMOS
total memory 256K bytes
power < 4 watts/chip
scalable to many chips
learning on chip
saturation arithmetic
bit jamming, virtual zero memory

Figure 1: Adaptive Solutions X1 (CNAPS) architecture

The chip is quite large - over 5 cm^2. Redundancy is built in to improve yield. Each chip has 80 processors, only 64 of which need to work. The X1 is implemented in 0.8 micron CMOS. It is expected to run at 25 MHz, and dissipate about 4 watts, for an energy per connection of about 2 nJ.

The chip requires an external instruction sequencer. In typical configurations, a single sequencer will control a number of chips.

The X1 should turn out to be a useful research tool in studying the behavior of learning algorithms in larger networks.

3.9 Waferscale implementations

Raffel et al from MIT Lincoln Labs [63] discuss a possible waferscale network using multiplying digital to analog converters (MDACs) using their Restructurable VLSI (RVLSI) approach to waferscale integration.

Yasunaga et al of Hitachi [80, 81] present a completely digital implementation which uses reduced precision parallel arithmetic. It is implemented using a 0.8 micron CMOS waferscale gate array. It has 540 neurons and 64 synapses per neuron. The wafer has 9 bit neural activations and 8 bit synaptic weights. Each weight w_{ij} is stored as a pair (j, w_{ij}) to efficiently represent sparse networks.

The paper reports a "step speed" of 464nsec, which appears to indicate the system clocks at 2.15 MHz. A unique j is generated on each step; only those neurons with matching (j, w_{ij}) pairs participate in the computation. If all the weights are utilized, 64×576 connections can be processed in 576 "steps", for a performance of 64 connections per step, or 138 MCPS.

3.10 Neurocomputer implementations

There have been several papers describing implementations of neural nets on massively parallel processors or neural coprocessors. We only mention those that have reported performance by way of comparison with neurochips.

Blelloch and Rosenberg [12] discuss mapping backpropagation onto the Connection Machine, reporting 3 MCUPS. Pomerleau et al [61] discuss mapping backpropagation onto the CMU Warp, reporting 17 MCUPS. Watanabe et al [75] discuss backpropagation on the NTT AAP-2, reporting 18 MCUPS.

De Groot and Parker [26] describe mapping backpropagation onto the Lawrence Livermore SPRINT. The machine has 10^6 synapses, 64 transputers, dissipates 100 watts, and achieves 12 MCPS. That's 8μJ/connection.

3.11 Neural processor capacity and performance summary

244

who	ref	type	tech	capacity (N, S)	speed CPS	power chip, syn	learning	scalability (multichip)
Caltech	[2]	CCD	2μ	256, 64K	5e8	-	no	no
Chiang	[14]	CCD	3μ	144, 2016	1.44e9	2W, 1.4nJ	no	no
USC	[48]	cap	2μ CMOS	25, 2525	-	-	no	no
Matsushita	[56]	analog	2.2μ BiCMOS	64, 768	7.7e7	-	no	no
ETANN	[37]	analog	1.0μ EEPROM	64, 8192	8e9	2W, 250 pJ	no	no
NET32K	[23]	D,A	0.9μ CMOS	1024, 32K	3e11	160mW, 500fJ	no	yes
CLC	[5]	D,A	1.2μ CMOS	32, 1000	1e9	1W,1nJ	yes	yes
Mitsubishi	[9]	analog	1.0u CMOS	125, 10K	1e11	1.5W, 15pJ	yes	yes
JPL	[18]	analog	2.0μ CMOS	32, 1000	1e9	-	no	no
X1	[27]	digital	0.8μ CMOS	64, 256K	1e9	4W, 4nJ	yes	yes
Ouali	[59]	digital	2.0μ CMOS	-	-	-	no	yes
Duranton	[17]	digital	1.6μ CMOS	-	8e8	-	yes	yes
Hirai	[35]	digital	1.2μ CMOS	6, 84	8.4e4	-	no	yes
DNNA	[42]	stochastic	-	32, 1024	2e8	-	no	yes
STONN	[78]	stochastic	-	20, 2000	5e8	-	no	yes
TInMANN	[15]	stochastic	2.0μ CMOS	1, ?	-	-	yes	yes
Hitachi	[81]	digital wsi	0.8μ CMOS	540, 34K	1.4e8	-	no	yes
CM1	[12]	mpp	-	-	3e6	-	yes	-
Warp	[61]	systolic	-	-	17e6	-	yes	-
AAP-2	[75]	array	-	-	18e6	-	yes	-
SPRINT	[26]	systolic	-	-, 1e6	12e6	100W, 8μJ	yes	-
neuron	-	-	-	1e11,1e15	1e16	1W, 0.1fJ	yes	yes

Table 1: Summary of existing implementations

Table 1 gives a summary of existing implementations. It is not meant to be exhaustive; it is meant to be a sampling of a wide variety of implementations. Analog, waferscale, neurocomputer, and biological entries are for comparison purposes. The two numbers in the *power* column are chip power dissipation and energy per connection.

The remainder of this chapter discusses issues related to building digital VLSI neural networks. Large networks can easily exceed the capabilities of the technology, so it is important to try to maximize network performance and capacity while minimizing power consumption. One of the messages we will try to convey is that digital systems have certain strengths which can be exploited as technology scales down into the submicron regime.

4 Architectural weapons

The performance of digital VLSI systems can be enhanced using a variety of techniques. These are: pipelining, precision, iteration, concurrency, regularity, and locality. Of these, precision may be particularly useful in neural net applications, and iteration may become more widespread as technology scales down.

4.1 Pipelining

Neural networks are well suited to deep pipelining because the latency of an individual unit is not nearly as important as the computation and communication throughput of the system.

If a system is too heavily piped, too much area is taken in latches, and the load on the clock is too large, increasing clock skew and reducing the net performance gain.

A system clock based on the propagation delay of a 4:2 adder provides a good balance between area and perforamnce. This is typically 1/4 of a RISC processor clock period in the same technology.

4.2 Precision

Precision can have a significant impact on area, power, and performance. The area of a multiplier is proportional to the square of the number of bits. A 32×32 bit parallel multiplier is 16 times the area of an 8×8 bit multiplier. At 8 or fewer bits, a multiplier is just an adder. A Booth encoder can cut the number of partial products in half and a 4:2 adder can accept up to 4 inputs.

Precision should be leveraged where it is available in the system. For example, an $N \times N$ bit multiply generates a $2N$ bit result. A Hebbian style learning algorithm could compute a weight adjustment to higher precision than is available and choose a weight value probabilistically. Using "probabilistic update", our Boltzmann machine can learn with as little as 2 bits of precision in the weight store.

4.3 Iteration

Iteration involves using an area-efficient arithmetic element and a local high speed clock to compute a single result in several clock cycles. Iteration reduces logic area by time multiplexing resources at a higher frequency than can be managed globally.

The conventional way to achieve high computational throughput is to implement a parallel arithmetic element clocked at the system clock rate. An alternate approach is to generate a local clock which is at least some multiple of the system clock rate, and then to implement only a fraction of the arithmetic element.

The basic idea behind iteration can be illustrated by the following example. Suppose we want to build a multiplier which can accept two inputs and output a product on every clock cycle. Using conventional techniques, we would have to use N^2 full adders and clock the data through the multiplier at the system clock rate. Suppose, however, that we could generate a local clock which was at least twice the frequency of the system clock. We could then implement the multiplier in half the area using $N^2/2$ fulladders. Iterative structures trade space for time.

Iteration will become more widespread as technology scales down and local clock rates scale up.

Mark Santoro implemented a 64X64-bit multiplier using an iterative 16X16-bit array [66, 67, 68]. He built the clock driver out of a 4:2 adder so it would be sure to track temperature and process variations. He achieved 400 MHz operation in 0.8μ CMOS.

Iterative structures, although highly area efficient, are likely to be higher power than unrolled structures. If unrolled, nodes only serve one function. Iteration essentially time multiplexes nodes. If the roles of the two nodes are unrelated, then their states are uncorrelated, and they may have a higher probability of changing state from iteration to iteration, dumping extra charge and thereby dissipating more power.

4.4 Concurrency

Concurrency is a widely used technique for increasing performance through parallelism. The degree of parallelism will increasingly be limited more by power density than by area. Neural networks are intrinsically highly parallel systems.

4.5 Regularity

Regularity in an architecture or algorithm permits a greater level of system complexity to be expressed with less design effort. Neural nets are composed of large numbers of similar elements, which should translate directly to VLSI implementation.

4.6 Locality

Local connections are cheaper than global ones. The energy required to transport information in CMOS VLSI is proportional to the distance the information travels. The available communication bandwidth is inversely proportional to wirelength, since the total available wirelength per unit area in a given technology is constant.

The energy required to switch a node is CV^2, where C is the capacitance of the node and V is the change in voltage. C is proportional to the area of the node, which for fixed width wires, is proportional to the length of a node.

5 Building blocks

A number of basic circuits and circuit techniques can be used to advantage in designing digital neural networks. This section is not intended to be a thorough presentation of digital logic design, as there are many excellent sources for this [54, 74, 77, 32]. Rather, we assume a familiarity with the basics and seek to highlight specific structures which are especially useful in designing digital neural network arithmetic elements, memory, and noise sources.

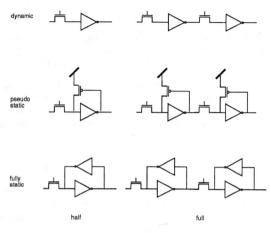

Figure 2: latches

5.1 Logic design styles

There are many logic design styles to choose from in implementing CMOS circuits. Logic design styles achieve different tradeoffs in speed, power, and area. The highest speed logic families also tend to consume the most power. The most compact tend to be slow.

Complementary pass-transistor logic (CPL) seems to offer modest performance, is compact, and low power. Yano et al's multiplier paper [79] has a good explanation of the style as well as detailed schematics of CPL arithmetic elements.

5.2 Latches and clocking

Latches play an important role in any digital design, especially if the design is pipelined. The right latch to use depends on the clocking discipline and the desired performance. Some clocking styles are safer than others. We have been using nonoverlapping two-phase clocks in our chips to date, but are using single phase clocking on our Boltmann chip.

Most of our designs use either the fully static or pseudo-static latch shown in Figure 2. There is a good section on latches in Weste and Eshraghian's book [77].

5.3 4:2 arithmetic

Carry propagation is an expensive operation in digital arithmetic. Several families of arithmetic have been developed to reduce the impact of carry propagation. Signed digit [10] and various redundant binary methods [28] have been proposed. We like "4:2" arithmetic based

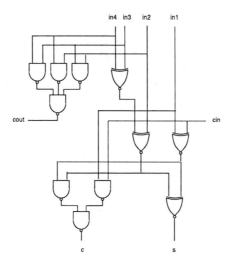

Figure 3: 4:2 adder

on 4:2 adders [70] because it interfaces cleanly to standard two's complement, and implements an efficient, compact accumulator [51, 50, 49, 66, 67, 68]. Although 4:2 adders can be implemented using two full adders, we discovered a "direct logic" implementation [51] (see Figure 3) that reduces the number of xors in series from four to three, increasing the speed by 33%.

5.4   Fast comparators

Unloaded manchester carry chains make very fast, efficient comparators. They are generally much faster than adders because there are no sum terms to load down the carry chain (see Figure 4). Comparators are needed in winer-take-all networks.

5.5   Memory

Synaptic storage density is an important issue in large scale networks, so memory optimization is important. One-transistor (1T) dynamic random access memories (DRAMs) have the highest density. Six-transistor (6T) static memories (SRAMs) consume the least power. A typical 1T DRAM cell measures $6\lambda \times 11\lambda$ $(66\lambda^2)$ [43]. A typical 4T SRAM measures $12\lambda \times 20\lambda$ $(240\lambda^2)$ [3]. Shift registers can often be implemented with either SRAM or DRAM, saving substantial amounts of area and power.

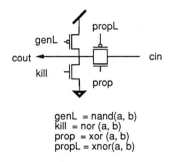

genL = nand(a, b)
kill = nor (a, b)
prop = xor (a, b)
propL = xnor(a, b)

Figure 4: unloaded manchester carry chain

5.5.1 DRAM

DRAM cells must be refreshed due to leakage current [13], and therefore consume more power than SRAMs. Normally the refresh power is a small fraction of the operating power, but could be significant in very large networks (see Section 10).

Figure 5 shows a variety of DRAM circuits. One-transistor (1T) DRAMs are the most compact but the most difficult to sense and control.

5.5.2 SRAM

Figure 6 shows a variety of SRAM circuits. Commercial SRAMs are normally implemented using a high resistance poly pullup, and can achieve densities only a factor of 4 worse than 1T DRAM.

5.5.3 Logic processes

The trench capacitors in DRAMs and high resistance poly pullups in SRAMs are not normally available in standard logic processes. A one-transistor DRAM was fabricated through MOSIS [22, 73], although the cell was quite large ($18.75\lambda \times 13.75\lambda$ ($258\lambda^2$).

5.5.4 Multi-level storage

Several researchers have reported techniques to store multiple levels in a single DRAM cell. The first were Heald and Hodges [30] in 1976. More recently, Aoki, Horiguchi, and colleagues [8, 39] reported 16 levels per cell in 1987. More levels might be achievable in neural networks since errors will always be small. Weight decay could be implemented by under-refreshing the memory.

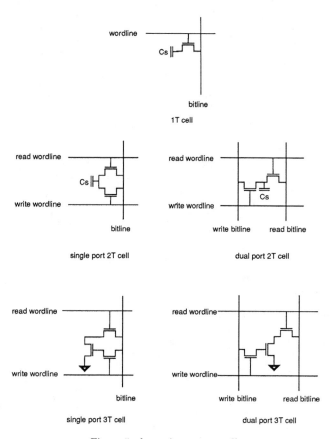

Figure 5: dynamic memory cells

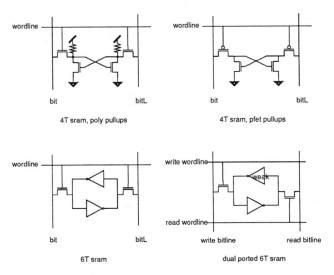

Figure 6: static memory cells

5.5.5 Sense amplifiers

Figure 7 shows a high performance differential dense amplifier reported in [16], and recommended for use with 1T DRAMs. This sense amp has the nice property that the crosscoupled inverters automatically perform a refresh write after read.

Figure 8 shows a current mirror sense amplifier similar to one reported in Aizaki et al [3] in 1990.

Figure 9 shows a single ended sense amp which can be used with ROMs, 3T DRAMs, and single-ended SRAMs.

5.6 Noise sources

Noise plays an important role in many neural network algorithms. Generating a large number of uncorrelated noise sources is a hard problem. Joshua Alspector and colleagues have developed an area-efficient technique which they have implemented on their CLS chip [5]. They give this subject a thorough treatment in [6].

5.7 Stochastic computation

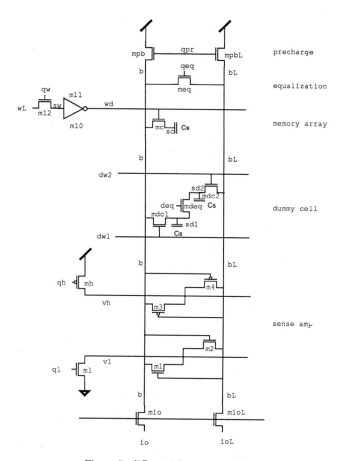

Figure 7: differential sense amplifier

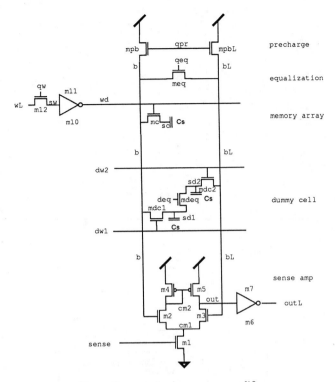

Figure 8: current mirror sense amplifier

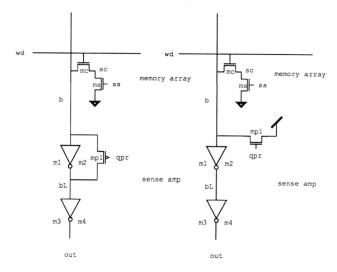

Figure 9: single ended sense amplifier

Stochastic computation implements multiplication and addition using probabilities. It is particularly effective when a computation can be spread out over many cycles. This is often the case in neural network learning, where the network evolves incrementally over many cycles. Stochastic techniques have been applied successfully to Hopfield networks [42], competitive learning networks [15], and Boltzmann machines [7, 5].

Figure 10 shows the two basic operations: multiplication by ANDing pulse streams ($P(AB) = P(A)P(B)$), and addition by ORing ($P(A + B) = P(A) + P(B) - P(AB)$). The $P(AB)$ term in the probabilistic addition limits the available dynamic range of the computation.

Although stochastic techniques are very efficient at implementing local computation, they are much less efficient communicating globally because the energy required to transmit data with a dynamic range of N is proportional to N rather than to $\log(N)$ as in standard digital encoding.

Good discussions of stochastic computation can be found in [19, 52, 20].

6   Area, performance, and power estimation

We have developed a simple area, performance, and power estimation technique which we use to construct spreadsheets in the early stages of architectural exploration, feasibility analysis, and optimization of a new design. Area is computed by estimating the number of transistors required. Performance is estimated by building an RC timing model of the critical paths into

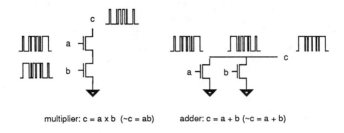

multiplier: c = a x b (~c = ab)   adder: c = a + b (~c = a + b)

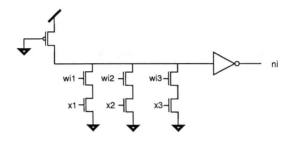

- and product
- wired-or sum-of-products

Figure 10: stochastic computing elements

the spreadsheet. Power is estimated using CV^2f, where f is obtained from the performance section of the spreadsheet.

Area, performance, and power are parametrized by technology. We have a "technology section" of the spreadsheet where we build in technology scaling rules to compute transistor transconductance and device parasitics.

6.1 Area estimation

We use a simple technique to estimate area of chips before we build them. We identify the major resources on the chip, and estimate the number of transistors for each resource. We then multiply the number of transistors in each case by an area-per-transistor which depends on how regular and compact we think we can make the layout. One-transistor DRAM and ROM are about $100\lambda^2$/xstr. 3T DRAM and 6T SRAM are about $200\lambda^2$/xstr. Tightly packed, carefully handcrafted logic is also about $200\lambda^2$/xstr. Loosely packed full custom logic is about $300\lambda^2$/xstr. Standard cells are about $1000\lambda^2$/xstr.

Block routing takes about 30% of the chip area. Standard cell routing takes 60% of the block. The pad frame reduces the die by about 1mm. For example, the largest die available on a standard MOSIS run is 7.9x9.2mm. Of this, 6.9x8.2mm, or 56.58mm^2, is available for logic and routing. Of this, 17mm^2 is routing, and 40mm^2 is logic. In 2 micron CMOS, lambda = 1 micron, so there is room for 133,000 transistors at $300\lambda^2$/xstr

The number of transistors required to implement a function can vary significantly depending on the design style. For example, a full adder implemented with gate logic requires about 30 transistors. However, it can be implemented in 15 transistors using pass-transistor logic. Which is best depends on desired performance and power dissipation, input drive and output load.

We maintain a list of leafcells, the number of transistors they require, and their area-per-transistor, which we reference in estimating requirements of new designs.

We also have a set of "tiling functions" which we use to construct complex blocks. For example, an $N \times M$ bit multiplier requires roughly NM full adders, whether it is implemented as an array or a tree.

This technique is especially well suited to spreadsheet implementation, and is especially useful during the early stages of architectural exploration and feasibility analysis in an area-limited design.

6.2 Performance estimation

We estimate performance using a simple RC timing model based on the RSIM simulator [72], in which transistors are calibrated to have an effective resistance charging or discharging a node capacitance.

We build the following equations into our spreadsheet models to allow the performance estimates to scale with technology. The symbology of these equations follows the development
in Hodges and Jackson [36].

AAA

```
Effective resistance of transistors

    Ilin = k/2(2 * (Vgs-Vt) * Vds - Vds^2)    Vds <= Vgs - Vt
    Isat = k/2(Vgs - Vt)^2                     Vds >= Vgs - Vt

    Iav  = integrate(I * dt)/T
    reff = DV / Iav
         = const * k

    kn   = un * cox      kp   = up * cox       A/V^2
    rn   = 1/kn/(Vdd - Vt)  rp  = 1/kp/(Vdd - Vt) ohms/sq
    Rn   = rn * l / w    Rp   = rp * l / w     ohms

Parasitic capacitance

    cg   = eox / tox                           farad/m^2
    cox  = cg                                  farad/m^2
    xj   = (2*esi/q/NA*(V-Vt))^.5              meters
    cj   = esi / xj                            farad/m^2
    cjsw = 3 * cj * xj = 3 * esi               farad/m
    ci   = eox / hi                            farad/m^2
    cisw = ci * ti                             farad/m

Propagation delay

    tpu  = Rp * (Cd + Ci + Cg)                 sec
    tpd  = Rn * (Cd + Ci + Cg)                 sec
```

6.3 Power estimation

There are three principal components to power dissipation in most CMOS systems:

$$P_{dc} = V^2/R_{dc}$$

$$P_{sc} = I_{sc}V$$

$$P_{ac} = CV^2f$$

where

> V is the supply voltage
> C is the total capacitance being switched
> f is the clock frequency
> R_{dc} is the total static pullup or pulldown resistance
> I_{sc} is the short-circuit current
> P_{dc} is the power dissipated at DC
> P_{sc} is the power dissipated due to short-circuit current
> P_{ac} is the power dissipated by switching capacitance.

P_{dc} can be designed out of a system, except leakage, which is usually on the order of a few microwatts [13, 29], but P_{sc} and P_{ac} cannot. In a CMOS inverter, I_sc is the current which flows when both the nfet and pfet are on during switching. Powell and Chau [62] have reported that P_{sc} can account for up to half the total power.

We have done some investigations which suggest that the short circuit current can be significant if rise times are long and transistors large, and that in most cases P_{sc} can be reduced to less than 10% of total power by sizing transistors. This implies that short circuit current can be more of a problem in gate array or standard cell design, where transistors are fixed sizes or are sized to drive large loads.

Consider a single CMOS inverter driving a purely capacitive load. Initially, assume the gate is at 0 volts, so the pfet is on, the nfet is off, and the output is at 5 volts. Now, switch the gate from 0 to 5 volts. Ideally, the pfet should turn off instantly, the nfet should turn on and drain the charge off the output (actually supply electrons to the output) until the output potential reaches 0 volts. The work done (or energy consumed) by the inverter is just QV where Q is the charge on the output and V is the initial potential difference between the output and GND. But $Q = CV$ so the work done is CV^2.

In practice, the input does not switch instantly, so both the nfet and the pfet are on for a short time, causing excess current to flow.

We measured the short circuit charge for a variety of transistor sizes, rise times, and output loads using spice [58] on a typical 2 micron CMOS technology from MOSIS.

Figure 11 shows the current flowing through vdd and gnd supplies as the gate is switched first from 0 to 5 volts between 2ns and 7ns, and then from 5 to 0 volts between 40ns and 45ns. The short circuit current in each case is the smaller spike; it is the current flowing through the supply which should be off. The short circuit charge is the area under the short circuit current. In the figure the short circuit charge is about 15% of the charge initially on the output, so P_{sc} will be about 15% of P_{ac}.

Figure 12 shows short circuit charge as a percentage of output load vs input rise/fall time. Each graph has a pair of curves for each of 5 output loads: 0, 100, 200, 500, and 1000 fF. One curve in each pair is for a rising input, the other for a falling input. The short circuit charge for a given output load is nearly the same whether the input is rising or falling. The pair with no output load has the largest short circuit charge.

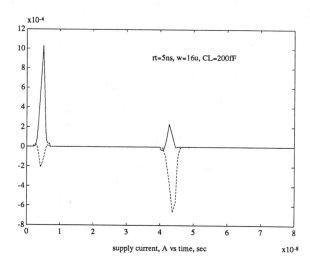

Figure 11: short circuit current

The short circuit charge is negative for fast rise times because the coupling capacitance between the gate and the fet drains pulls the output above 5 volts. This deposits additional charge on the output which must then be removed. So there is an energy cost associated with switching an input too fast, and a finite rise time at which the net short circuit charge is exactly zero.

We want to investigate this phenomenon in more detail, and see if we can modify our transistor sizer to take advantage of it. We also want to extend our analysis to complex gates, and other logic design styles.

In practice, we estimate power in a variety of ways. In some cases, when we have experience or know statistically what percentage of the nodes are switching, we use CV^2f. In other cases, we use Powell and Chau's power factor approximation (PFA) technique, which uses the energy of existing devices to predict new ones. In still other cases, when the devices have an analog behavior, such as sense amplifiers in memories and reduced voltage swing logic, we use spice to compute current and integrate to find charge dumped.

We have modified the RSIM timing simulator to accumulate the charge dumped as nodes switch during simulation. This is easy to do if the simulator is event driven. We compared RSIM's results on a signal processing chip we fabricated through MOSIS [40] with power measurements done on a performance tester and found agreement to within 20%.

6.4 Technology scaling rules and examples

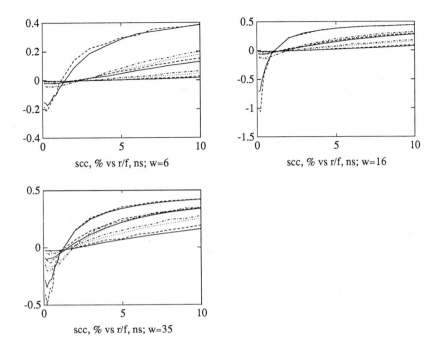

Figure 12: short circuit charge as a percentage of output load vs rise time for different size devices

param	scaling	description
tech	S	
tox	$1/S$	gate oxide
cox	S	gate capacitance
uo	1	mobility
k	S	transconductance
r	$1/S/V$	resistance
xj	$V^{1/2}$	junction depth
cj	$1/V^{1/2}$	diffusion capacitance
cjsw	1	diffusion sidewall capacitance
hi	$1/S$	metal elevation
ti	$1/S$	metal thickness
ci	S	interconnect area capacitance
cisw	1	interconnect sidewall capacitance
R	$1/S/V$	device resistance
C	$1/S$	device capacitance
tp	$1/S^2/V$	propagation delay
I	SV^2	current
i	S^3V^2	current density
p	SV^3	power per device
P	S^3V^3	chip power dissipation

AAA

Table 2: Technology scaling

tech		2.0	1.6	1.2	1.0	0.8	0.6	
S		1.0	1.3	1.7	2.0	2.5	3.3	
S^2		1.0	1.7	2.9	4.0	6.3	11.0	
clk	S^2	20	34	58	80	125	220	MHz
clk	S^2	80	136	232	320	500	880	MHz
R(12:2)	$1/S$	6.0	4.6	3.5	3.0	2.4	1.8	kohm
Cg	$1/S$	20.0	15.4	11.8	10.0	8.0	6.1	ff
Cd	$1/S^2$	40.0	23.5	13.8	10.0	6.3	3.6	ff
Ci	$1/S$	40.0	30.8	23.6	20.0	16.0	12.2	ff
tgate	$1/S^2$	1.38	0.81	0.48	0.35	0.23	0.13	nsec
t4:2	$1/S^2$	12.5	7.35	4.31	3.13	1.98	1.14	nsec

Table 3: scaling example

To scale area, performance, and power to a desired technology, we build into our spreadsheet the parameters of a base technology (2.0 micron CMOS in our case; $\lambda_0 = 1.0\mu$), compute $S = \lambda_0/\lambda$, and apply the equations in table 2. We leave V, the supply voltage, explicit so we can compare the impact of "constant voltage" scaling and full scaling. These equations follow the development in Hodges and Jackson [36].

Table 3 shows how actual 2μ numbers would look scaled for a range of technologies. In the table, R(12:2) is the effective resistance of a $12\lambda \times 2\lambda$ transistor. Cg, Cd, and Ci are nominal gate, diffusion, and wiring capacitances. tgate is the propagation delay of a single gate. t4:2 is the propagation delay of a 4:2 adder.

Table 4 shows typical spice parameters for different technologies. The "exp" row in the table shows the exponent which would be supplied to spice. For example, for 2.0μ, tox=403e-10. cpa, $m1a$, and $m2a$ are area capacitance of polysilicon, metal1, and metal2 in farads/meter^2.

The information for 2.0, 1.6, and 1.2 μ technologies was obtained from MOSIS. 16 runs were averaged in each technology, and the deck closest to the mean was selected as nominal. The 0.8 micron information was obtained from a much smaller dataset.

These technologies all run at 5 volts, so S^2 performance is predicted by the formulas in Table 2. However, Table 4 shows that mobility, uo, is not constant. Comparing 2.0μ and 0.8μ, uo is scaling as $1/S^{1/2}$. Now R scales as $1/S/V/uo = 1/S^{1/2}/V$ and C as $1/S$ so propagation delay tp should then scale as $1/S^{3/2}/V$.

According to Hodges and Jackson, mobility is a function of substrate doping, N_A. N_A

tech	u0n	u0p	cgdon	cgdop	cjswn	cjswp	xjn	xjp	nsubn	nsubp
exp	0	0	-12	-12	-12	-12	-9	-9	+15	+15
2.0	631	237	298	285	548	334	250	50	5.76	6.24
1.6	583	186	573	494	588	184				
1.2	574	181	628	324	423	159				
0.8	447	101	229	271	200	200	157	138	85.58	79.65

tech	tox	cpa	cjn	cjp	m1a	m2a		
exp	-10	-6	-6	-6	-6	-6	tox	$1/S$
							cpa	S
2.0	403	388	130	262	26	19	cjn	$S^{3/2}$
1.6	250	573	140	432	35	23	cjp	S
1.2	209	644	293	481	36	24	m1a	S
0.8	170	794	547	570	79	31	m2a	$S^{1/2}$

Table 4: nominal process parameters for different technologies

increases as λ decreases. As N_A increases, collisions become more likely so mobility decreases. If the increase in N_A is necessary to prevent breakdown in the presence of higher E fields in constant voltage scaling, then presumably NA could be kept constant if V were scaled.

Table 5 shows predicted performance of Mark Santoro's self-timed 4:2 adder based clock driver [68] in various technologies based on both constant V ($S^{3/2}$) and scaled V (S). The numbers are in good agreement with observed performance at 2.0, 1.6, and 0.8μ CMOS. Extrapolations below 0.5μ are highly speculative. They serve as an upper bound on expected performance.

Process variations can impact performance significantly. We have found the ratio of "fast-fast" models to "slow-slow" models to be a factor of two in performance, with "typical" models right in the middle.

$$ff/typ = typ/ss = \sqrt{2}$$
$$ff/ss = 2$$

Performance degrades by a factor of 1.8 from 25degC, 5V to 130degC, 4.5V. Worst case performance can therefore be as much as four times slower than best case when process and temperature variations are combined.

7 Design techniques for maximizing performance

tech	tpd	MHz constV	Vdd	MHz scaleV
2.00	10.83n	92	5	
1.60	7.58n	132	5	
1.40	6.12n	163	5	
1.20	4.78n	209	5	
1.00	3.57n	280	5	
0.80	2.50n	400	5	
0.60	1.58n	634	3.3	418
0.50	1.18n	849	3.3	560
0.40	825p	1213	3.3	801
0.30	520p	1921	3	1153
0.25	389p	2572	3	1543
0.10	90p	11143	2	4457

Table 5: 4:2 adder-based clock circuit performance

	const V	scale V
devices/area	S^2	S^2
speed	S^2	S
ops/sec/area	S^4	S^3
energy/op	$1/S$	$1/S^3$
power/area	S^3	1

Table 6: Energy scaling

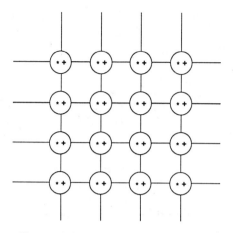

Figure 13: SP: one processor per synapse

For many problems in signal processing, performance can be maximized by maximizing pipelining and parallelism. This is an excellent technique for small and medium scale systems with generous power budgets. But as systems get very large, the level of parallelism achievable either by pipelining or replication will be limited by power considerations. Table 6 suggests that scaled voltage is the way to go for implementing very large networks.

In highly pipelined systems, performance is maximized by placing pipe stages so delay is the same in every stage. This requires good timing analysis and optimization tools.

As technology scales, the clock rates achievable by highly pipelined structures will be difficult to distribute globally. Self-timed iterative circuits can save area by performing high speed local computation.

8 Technology scaling of VLSI neural networks

We have applied the area and performance estimation techniques described in Section 6 to three different inner product processor architectures. The first (SP; see Figure 13) has one processor per synapse. The second (NP; see Figure 14) has one processor per neuron, with the synaptic weights stored in memory local to each processor, similar to the Adaptive Solutions X1 architecture. The third (FP; see Figure 15) has a fixed number of processors on chip, and off-chip weights.

SP has the lowest synaptic storage density but the highest computational throughput. FP has the lowest throughput, but the highest synaptic storage density and unlimited capacity.

The number of processors which can be placed on a single FP is I/O limited. Assuming 4

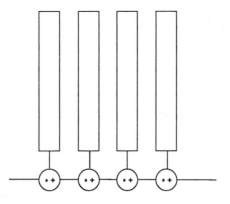

Figure 14: NP: one processor per neuron

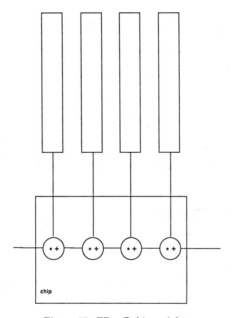

Figure 15: FP: off-chip weights

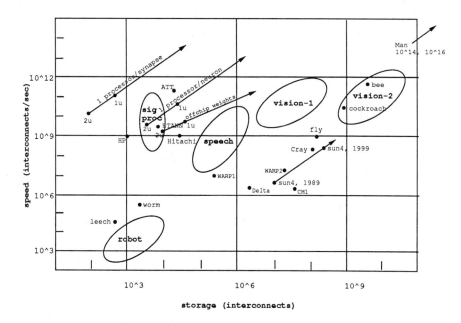

Figure 16: technology and performance scaling

bit weights, 128 pins would be required to support 32 processors. The I/O constraint can be substantially alleviated with multichip module (MCM) packaging [41]; this allows far more flexibility in choosing die size and system partitioning.

8.1 Technology and performance scaling of inner product processors

In Figure 16 we plotted the technology scaling trendlines of the SP, NP, and FP architectures on the DARPA-style capacity-vs-performance graph, along with the DARPA application requirements.

In Figure 17, we plotted the waferscale integration trendlines of the three architectures. The FP architecture has two trendlines. Since its weights are offchip, a waferscale implementation has the option of replicating memory alone. In this case performance remains constant, but capacity improves. Interestingly, of the options shown, this one most closely matches the requirements of the DARPA applications.

In Figure 18, we derive the architecture from the application and the technology. In a very rough sense, an application can be characterized by its capacity and performance re-

268

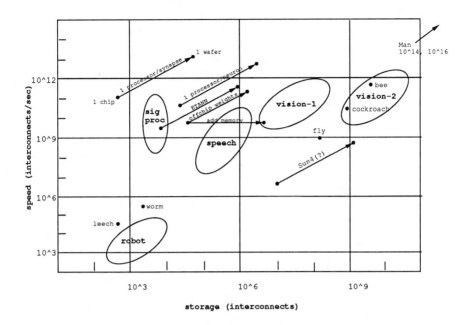

Figure 17: waferscale integration at 1.0μ

269

Applications and architectures

$$\text{connections/processor} = \frac{\text{connections/system}}{\text{connections/sec/system}} \times \text{connections/sec/proc}$$

(architecture)	(application)	(technology)

Application	connections	cps	connections/cps	cpp 2.0u	cpp 1.0u	cpp 0.6u
robot arm	3e3	3e4	1e-1	8e6	2e7	4e7
signal proc	3e4	1e10	3e-6	2.4	600	1500
speech	1e6	1e9	1e-3	8e4	2e5	5e5
vision-1	1e8	1e11	1e-3	8e4	2e5	5e5
vision-2	1e10	1e11	1e-1	8e6	2e7	5e7

Technology capacity and performance

Tech	Clock(S^1.5)	chip 3TDRAM	chip 1TDRAM	wafer 3TDRAM	wafer 1TDRAM
2.0	80 MHz	1.5e3	4.5e3	1.5e5	4.5e5
1.0	200 MHz	6.0e4	1.8e5	6.0e6	1.8e7
0.6	500 MHz	2.0e5	6.0e5	2.0e7	6.0e7

Application	Nprocessors	Nchips (wafers)
robot arm	e-4	e-2 chips
signal proc	50	1 chip
speech	5	10 chips
vision-1	500	10 wafers
vision-2	500	1000 wafers

Figure 18: algorithms and architectures

quirements. An architecture can be loosely defined in terms of the connections serviced by a processor. For example, the SP architecture has one connection per processor. The technology defines the available performance. According to Figure 18, the product of the available performance and the ratio of the desired capacity and performance determines the architecture.

Figure 18 suggests that SP is most appropriate for the signal processing application in 2.0 micron CMOS, but that NP is more appropriate in 1.0 and 0.6 micron. By comparing chip DRAM capacities in the table (5 bit weights), we see that the FP architecture is best in all other cases except speech and vision-1 in 0.6 micron, where NP is best.

9 The Stanford Boltzmann Engine

Boltzmann machines [33, 1, 7, 34, 60] are a special class of neural networks whose learning algorithm can be shown to minimize a global energy measure using only local information. The network contains recurrent connections (feedback), which can result in multiple responses to a single stimulus depending on the time evolution of the neural activations in the presence of noise.

Alspector has shown that both the deterministic (mean field) and stochastic variations of the Boltzmann learning algorithm can learn NETtalk [69] as well as the backpropagation algorithm. He says the computational effort of mean field learning is $3\times$ backprop; stochastic is $100\times$.

Three Boltzmann chips have been implemented to date [7, 9, 5]. Each uses an analog current sum and adaptive synaptic processors.

We are implementing a fully digital Boltzmann engine (see Figure 19) which stores synaptic values and neural activations in one-transistor dynamic RAM and implements the $w_{ij}x_j$ products with 64 pipelined 5×5-bit multiplier accumulators operating in parallel. We are designing for 80 MHz operation, 100 pJ per connection in 2 micron CMOS, and a feedforward computation rate of 5 GCPS.

During Boltzmann learning, the chip will complete one anneal step every 800 nsec. With 32 temperatures in the anneal schedule (we have found that an N neuron network needs about $N/2$ temperatures in the anneal schedule), the chip will process one training vector every 25 μsec, and learn to identify it in 100 to 1000 presentations. Because the implementation is purely digital, previously learned weights can be up- and down-loaded between the chip and a host.

Our simulations indicate 5 bits of precision are sufficient for both neural activations and synaptic weights, and that as few as 3 bits can be used with "probabilistic update", wherein an available weight value is selected with probability proportional to a higher precision accumulated inner product.

We plan to implement a single sigmoid unit as a lookup table. We plan to implement the temperature multiplier using reduced precision floating point with a 2 bit mantissa and a 4

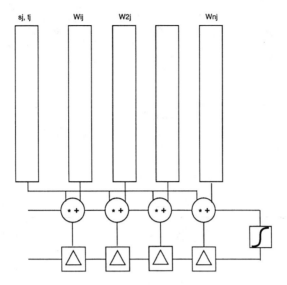

Figure 19: the Stanford Boltzmann Engine

bit exponent.

The multiplier-accumulators accumulate in carry-save form using 4:2 adders to reduce hardware and increase performance. To avoid delivering data to the sigmoid unit in carry-save form, a few extra cycles are appended to the accumulation to flush the carries.

There are two weight update processors, since N^2 weights need to be updated every $N^2/2$ cycles (N cycles per inner product product $\times N/2$ temperatures).

10   Large networks

VLSI faces formidable obstacles implementing large neural networks. Although the networks are massively parallel, there is evidence that communication is predominantly local [11, 64, 53]. Furthermore, neurons are slow compared to VLSI, assuming a simple neuron model, implying that computation and communication should be time multiplexed. Digital implementations lend themselves well to time multiplexing, but this requires a projection or folding of the network onto the hardware.

Rudnick and Hammerstrom of the Oregon Graduate Center Cognitive Architecture Project wrote an excellent paper on a waferscale architecture for large scale neural networks [64]. They describe a waferscale communication architecture involving dual concentrate/broadcast domain trees. They assume that most communication is local, and that when a neural activation changes, it is broadcast to the smallest domain that completely contains its fanout. Most often, this is just its nearest neighbor, but it could be the entire network.

10.1 Energy/power

Carver Mead wrote an excellent paper for the October 1990 Proceedings of the IEEE [53], in which he discusses the energy requirements of electrical and biological systems. He makes the point that in systems on the scale of human cortex, with 10^{16} synapses, the energy budget is a primary constraint. The human brain consumes a few watts, so each synaptic computation is constrained to require on the order of 0.1fJ.

In 2μ CMOS, a single minimum size (4×2) transistor is 7fF. That translates to 175fJ at 5V. C_g scales as $1/S$. In 0.5μ CMOS, a single minimum size transistor is 1.75fF, or 16fJ at 3V. Projecting a few years down the road to 0.1μ technology, and 1V supplies, a single transistor might consume 0.350fJ.

Mead maintains that the energy required to switch a single transistor is comparable to the energy to perform one complex synaptic computation in a biological network, and that digital systems switch about 10000 transistors to perform a single operation, making them 10000 times less efficient than biological systems.

A 5x5 bit multiplier-accumulator can be built using roughly 1000 transistors such that only about 100 switch during a computation, so we might hope to get within a factor of 100 of biology using reduced precision digital arithmetic.

There is also an energy cost associated with retrieving synaptic weights from a memory structure. If the synaptic store is divided into 64x64 bit blocks, only one of which is accessed, about 100 transistors will switch retrieving a weight from the memory.

There is also an energy cost associated with the sigmoid, but it is insignificant since it only happens once for every 10,000 synaptic computations.

There is also an energy cost associated with delivering an activation to its synaptic destinations. This is probably a small percentage of the synaptic computation cost if connectivity is predominantly local, and each processor handles a large number of neurons.

There is also an energy cost associated with learning. Hebbian learning [31], $dw_{ij} = \epsilon s_i s_j$, could be implemented with a single multiplier, and could be kept to a fraction of the synaptic computation cost by slow adaptation.

So it looks like synaptic computation and weight retrieval dominate the energy cost in a digital network, and that this cost can be brought within a factor of a few hundred of biology (see Figure 20).

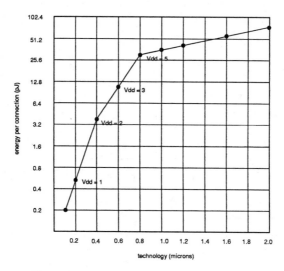

Figure 20: energy per connection vs technology

10.2 Sparsely connected nets

If the network topology is not physically expressed in the hardware, then it must be folded onto the hardware topology. This folding results in time multiplexed utilization of resources, and requires labeling of interconnections. Unless the network topology can be reconstructed algorithmically, data flowing in the system will have to be labeled. Another way of saying this is that as soon as the network becomes sparse, state information in the network (neural activations and synaptic weights) must be accompanied by an address. In the limit, this would increase memory storage requirements by a factor of 37 (since $2^{37} = 10^{11}$), which, though onerous, is far better than implementing a fully connected network, which would increase memory storage requirements by a factor of 10^6.

Algorithmic and explicit labeling can be combined by partitioning the network into a hierarchy of subnetworks and recursively labeling the subnets down to some level at which algorithmic labeling is introduced. For example, if the average neural fanin is 10,000, the network could be partitioned into fully connected sets of 10^4 neurons, with 10^8 synaptic weights in a set.

If we can avoid labeling synapses, and just label neural outputs, then communication bandwidth increases by a factor of 37, but synaptic storage stays the same. This is fortunate, since communication bandwidth is the easiest to increase. We can do this by grouping neurons in fully connected clusters. There will be a slight increase in synaptic storage required since the neurons will not in general be fully connected.

The most straightforward way to specify the interconnections among N neurons is with a matrix of N^2 elements. Suppose the network connections are random and uniformly distributed, ie the probability that any two neurons are connected is a constant P. Then the probability that a randomly selected matrix element is nonzero is also P, and the total number of nonzero weights is PN^2. If P is less than $log_2(w_{max})/(log_2(w_{max}) + log_2(N))$, it is more efficient to store the address of each neuron along with each weight than to store all the weights. That is, rather than store $(w_{ij} : i = 1, N; j = 1, N)$, store $(w_{ij}, j : w_{ij} <> 0)$. It is not necessary to store i explicitly if weights are clustered by output. However, the input neuron ids must be stored along with each weight if the network is randomly connected.

Any regularity which might exist in the pattern of network connections can be exploited to reduce the number of neuron ids which must be stored. In some cases, such as fully connected networks, or networks whose pattern of connectivity is the same for every neuron, id storage can be eliminated, since weights can be allocated to processors in a completely uniform way.

If any subset of the network has a regular connection pattern, then id storage for the neurons in that subnet can be eliminated. For example. in a multilayer feedforward net, if each pair of adjacent layers is fully connected, then only one id is needed in each layer to identify the entire "block" of connections. All the connections within that block can be computed as a fully connected set relative to a single base address.

The number of block ids which must be stored is equal to the number of tiles required to cover the nonzero entries in the weight matrix. There is a significant tradeoff in tiling function complexity and flexibility. The more powerful the tiling function, the higher its computational cost. The objective is to find a set of functions and a tiling of those functions which maximizes coverage and minimizes overlap, at modest computational cost.

10.3 Biological complexity

VLSI may have been oversold in its ability to provide sufficient resources for large scale network computation. It is tempting to endow VLSI with essentially limitless capabilities, so that any problem, no matter how intractable, can be solved. In neural networks, VLSI may have met its match. The major challenges are computation, communication, and memory. Since silicon is much faster, computation, and communication can be time multiplexed. Memory, however, cannot.

Table 7 gives some published estimates of performance and capacity of the human cortex. Tables 8 and 9 further refine those estimates into peripheral and central nervous system operations.

The bandwidth of a single metal wire is somewhere between 10 Mbit/sec and 1 Gbit/sec, depending on the technology. In the peripheral nervous system (PNS), the average bandwidth of a single nerve fiber is somewhere around 1 Kbit/sec. Therefore each silicon wire could carry the information of roughly 10^5 PNS neurons. In the central nervous system (CNS), the average bandwidth of a single axon is around 1 bit/sec, so each silicon wire could carry

	neurons	APS	synapses	CPS
[53]	-	-	10^{16}	10^{16}
[76]	10^{11}	10^{13}	10^{14}	10^{16}
Sejnowski PNS	10^9	10^{12}	10^{13}	10^{16}
Sejnowski CNS	10^{11}	10^{11}	10^{15}	10^{15}

Table 7: Cortical statistics

10^9	neurons
10^3	activations/sec/neuron
10^4	fanout
10^{13}	synapses
10^{12}	activations/sec
10^{16}	connections/sec

Table 8: Peripheral nervous system capacity and performance

10^{11}	neurons
1	activation/sec/neuron
10^4	fanout
10^{15}	synapses
10^{11}	activations/sec
10^{15}	connections/sec

Table 9: Central nervous system capacity and performance

	PNS	CNS
processors	10^7	10^6
neurons/proc	10^2	10^5
synapses/proc	10^6	10^9

Table 10: Cortical processor architectures

tech	syn/cm^2 S^2	macs/cm^2 S^2	CPS/cm^2 S^3	cm^2/10^{15}syn S^2	cm^2/10^{16}CPS S^3
2.00μ	1e6	1e2	1e10	1e9	1e6
1.00μ	4e6	4e2	8e10	2e8	1e5
0.50μ	2e7	2e3	6e11	4e7	1e4
0.25μ	1e8	6e3	5e12	1e7	2e3
0.10μ	1e9	4e4	8e13	1e6	1e2

Table 11: Capacity and performance (scale V)

the information of roughly 10^8 CNS neurons.

The computational bandwidth of a neuron is much more difficult to quantify because it depends to an overwhelming extent on the logical complexity of the model used to describe it. The secret is to find a good match between the set of available logic elements and the algorithm to be implemented.

Table 10 suggests that a single VLSI processor could conceivably process 10^9 synapses per second, and might therefore handle 10^2 PNS neurons or 10^5 CNS neurons. 10^7 PNS processors and 10^6 CNS processors would be required. Each PNS processor would service 10^6 synapses; each CNS processor 10^9 synapses.

Table 11 suggests that the synaptic storage problem is four orders of magnitude more challenging than achieving adequate performance. It also suggests that achieving comparable computational bandwidth is perhaps feasible, but that comparable storage capacity will require a technological breakthrough. The capacity numbers assume only one storage cell is required per synapse. Analog synaptic cells reported so far are two orders of magnitude larger (see Table 12).

type		$\lambda \times \lambda$	λ^2	λ^2/T	rel.area	who
1T	EPROM	8×8	64	64	1	commercial
1T	DRAM	8×8	64	64	1	commercial
4T	SRAM	12×20	240	60	4	commercial
1T	DRAM	8×16	128	128	2	MOSIS
2T	DRAM	13×31	403	200	6	MOSIS
3T	DRAM	24×24	576	192	9	MOSIS
6T	SRAM	32×36	1152	192	18	MOSIS
7T	synapse	40×60	2400	342	37	[48]
??T	ETANN	83×97	8036	??	125	[37]
42T	synapse	200×200	40000	952	625	[9]
200T	CLC	160×320	51200	256	700	[5]

Table 12: Area of various synaptic cells

The size of an individual synaptic storage element is the major limiting factor in determining the size of a network.

Table 12 illustrates the wide range in synaptic densities among reported alternatives.

Until someone comes up with a much more area efficient analog synaptic cell, or until technology scales well below 0.1 micron, adaptive analog synapses will remain too big to scale up to biology. The only VLSI chance for large nets is 1T DRAM or 4T SRAM.

Leakage current, according to Chatterjee et al[13], is $3fA/\mu^2$, and is a strong function of temperature. A 0.5μ 1T DRAM storage node with an area of $2\mu^2$ has a leakage current of 6fA/bit. That's a total leakage current of 6 amps for 1e15 bits.

An important consequence of these observations is that neurochip architectures which emphasize computational bandwidth but have low synaptic storage densities will not scale well to large networks.

11 Conclusion

Digital VLSI can be used to implement a wide variety of neural networks. Both very high performance and very large scale networks can be implemented effectively, especially as technology scales down into the submicron regime. There is a tradeoff between performance and flexibility. Two significant challenges for future research are increasing synaptic storage capacity and minimizing power consumption.

The connectionist approach assumes synaptic communication can be modeled simply. The limit to the size of networks which VLSI can implement depends not only on the energy per computation and the synaptic storage density, but also on the complexity of the synaptic interaction and the neuron model.

Acknowledgements

This work was supported by the NASA Center for Aeronautics and Space Information Systems (CASIS) under grant NAGW 419.

References

[1] David H. Ackley and Geoffrey E. Hinton. A learning algorithm for Boltzmann Machines. *Cognitive Science*, 9:147–169, 1985.

[2] Aharon J. Agranat, Charles F. Neugebauer, and Amnon Yariv. A CCD based neural network integrated circuit with 64K analog programmable synapses". In *IJCNN International Joint Conference on Neural Networks*, pages II:551–555, 1990.

[3] Shingo Aizaki, Masayoshi Ohkawa, Akane Aizaki, Yasushi Okuyama, Isao Sasaki, Toshiyuki Shimizu, Kazuhiko Abe, Manabu Ando, and Osamu Kudoh. A 15ns 4Mb CMOS SRAM. In *IEEE International Solid-State Circuits Conference*, pages 126–127, 1990.

[4] Yutaka Akiyama. *The Gaussian Machine: a stochastic, continuous neural network model.* PhD thesis, Keio University, February 1990.

[5] Joshua Alspector. CLC - A cascadeable learning chip. NIPS90 VLSI Workshop, December 1990.

[6] Joshua Alspector, Joel W. Gannett, Stuart Haber, Michael B. Parker, and Robert Chu. A VLSI-efficient technique for generating multiple uncorrelated noise sources and its application to stochastic neural networks. IEEE Transactions on Circuits and Systems, 1990.

[7] Joshua Alspector, Bhusan Gupta, and Robert B. Allen. Performance of a stochastic learning microchip. In *Advances in Neural Information Processing Systems*, 1988.

[8] Masakazu Aoki, Yoshinobu Nakagome, Masahi Horiguchi, Shin'ichi Ikenaga, and Katsuhiro Shimohigashi. A 16-level/cell dynamic memory. *IEEE Journal of Solid-State Circuits*, pages 297–299, April 1987.

[9] Yutaka Arima, Koichiro Mashiko, Keisuke Okada, Tsuyoshi Yamada, Atushi Maeda, Harufusa Kondoh, and Shinpei Kayano. A self-learning neural network chip with 125

neurons and 10K self-organization synapses. In *Symposium on VLSI Circuits*, pages 63–64, 1990.

[10] Algirdas Avizienis. Signed-digit number representations for fast parallel arithmetic. *IRE Transactions on Electronic Computers*, pages 389–400, September 1961.

[11] Jim Bailey and Dan Hammerstrom. Why VLSI implementations of associative VLCNs require connection multiplexing. In *IEEE International Conference on Neural Networks*, pages II:173–180, 1988.

[12] G. Blelloch and C. R. Rosenberg. Network learning on the Connection Machine. In *Proceedings of the Tenth International Joint Conference on Artificial Intelligence*, pages 323–326, 1987. Milan, Italy.

[13] Pallab K. Chatterjee, Geoffrey W. Taylor, Al F. Tasch, and Horng-Sen Fu. Leakage studies in high-density dynamic MOS memory devices. *IEEE Transactions on Electron Devices*, pages 564–575, April 1979.

[14] Alice Chiang, Robert Mountain, James Reinold, Jeoffrey LaFranchise, James Gregory, and George Lincoln. A programmable CCD signal processor. In *IEEE International Solid-State Circuits Conference*, pages 146–147, 1990.

[15] David Van den Bout and Thomas K. Miller III. TInMANN: The integer Markovian artificial neural network. In *IJCNN International Joint Conference on Neural Networks*, pages II:205–211, 1989.

[16] Sang H. Dhong, Nicky Chau-Chau Lu, Wei Hwang, and Stephen A. Parke. High-speed sensing scheme for CMOS DRAM's. *IEEE Journal of Solid-State Circuits*, pages 34–40, February 1988.

[17] M. Duranton and J.A. Sirat. Learning on VLSI: A general purpose digital neurochip. In *IJCNN International Joint Conference on Neural Networks*, 1989.

[18] Silvio Eberhardt, Tuan Duong, and Anil Thakoor. Design of parallel hardware neural network systems from custom analog VLSI 'building block' chips. In *IJCNN International Joint Conference on Neural Networks*, pages II:183–190, 1989.

[19] Brian R. Gaines. Stochastic computing. In *Spring Joint Computer Conference*, pages 149–156, 1967.

[20] Brian R. Gaines. Uncertainty as a foundation of computational power in neural networks. In *IEEE First International Conference on Neural Networks*, pages III:51–57, 1987.

[21] Simon C. J. Garth. A chipset for high speed simulation of neural network systems. In *IEEE First International Conference on Neural Networks*, pages III:443–452, 1987.

[22] James A. Gasbarro and Mark A. Horowitz. A single-chip, functional tester for VLSI circuits. In *IEEE International Solid-State Circuits Conference*, pages 84–85, 1990.

[23] Hans Peter Graf and Don Henderson. A reconfigurable CMOS neural network. In *IEEE International Solid-State Circuits Conference*, pages 144–145, 1990.

[24] H.P. Graf, W. Hubbard, L.D. Jackel, and P.G.N. deVegvar. A CMOS associative memory chip. In *IEEE First International Conference on Neural Networks*, pages III:461–468, 1987.

[25] H.P. Graf, L.D. Jackel, R.E. Howard, B. Straughn, J.S. Denker, W. Hubbard, D.M. Tennant, and D. Schwartz. VLSI implementation of a neural network memory with several hundreds of neurons. In *Neural Networks for Computing, Snowbird, Utah 1986*, pages 182–187. American Institute of Physics, 1986.

[26] A. J. De Groot and S. R. Parker. Systolic implementation of neural networks. In *SPIE, High Speed Computing II*, volume 1058, January 1989.

[27] Dan Hammerstrom. A VLSI architecture for high-performance, low-cost, on-chip learning. In *IJCNN International Joint Conference on Neural Networks*, pages II:537–544, 1990.

[28] Yoshihisa Harata, Yoshio Nakamura, Hiroshi Nagase, Mitsuharu Takigawa, and Naofumi Takagi. A high-speed multiplier using a redundant binary adder tree. *IEEE Journal of Solid-State Circuits*, pages 28–34, February 1987.

[29] Shigeyuki Hayakawa, Masakazu Kakumu, Hideki Takeuchi, Katsuhiko Sato, Takayuki Ohtani, Takeshi Yoshida, Takeo Nakayama, Shigeru Morita, Masaaki Kinugawa, Kenji Maeguchi, Kiyofumi Ochii, Jun'ichi Matsunaga, Akira Aono, Kazuhiro Noguchi, and Tetsuya Asami. A 1uA retention 4Mb SRAM with a thin-film-transistor load cell. In *IEEE International Solid-State Circuits Conference*, pages 128–129, 1990.

[30] Raymond A. Heald and David A. Hodges. Multilevel random-access memory using one transistor per cell. *IEEE Journal of Solid-State Circuits*, pages 519–528, August 1976.

[31] Donald O. Hebb. *The Organization of Behavior*. Wiley, 1949.

[32] John L. Hennessy and David A. Patterson. *Computer Architecture A Quantitative Approach*. Morgan Kaufmann Publishers, 1989.

[33] G.E. Hinton, T.J. Sejnowski, and D.H. Ackley. Boltzmann Machines: Constraint satisfaction networks that learn. Technical Report CMU-CS-84-119, Carnegie-Mellon University, May 1984.

[34] Geoffrey E. Hinton. Deterministic Boltzmann learning performs steepest descent in weight-space. *Neural Computation*, 1:143–150, 1989.

[35] Yuzo Hirai, Katsuhiro Kamada, Minoru Yamada, and Mitsuo Ooyama. A digital neurochip with unlimited connectability for large scale neural networks. In *IJCNN International Joint Conference on Neural Networks*, pages II:163–169, 1989.

[36] David A. Hodges and Horace G. Jackson. *Analysis and Design of Digital Integrated Circuits*. McGraw-Hill, 1983.

281

[37] Mark Holler, Simon Tam, Hernan Castro, and Ronald Benson. An electrically trainable artificial neural network (ETANN) with 10240 "floating gate" synapses. In *IJCNN International Joint Conference on Neural Networks*, pages II:191–196, 1989.

[38] J.J. Hopfield. Neural networks and physical systems with emergent collective computational properties. *Proceedings of the National Academy of Sciences, USA*, 79:2554–2558, April 1982.

[39] Masahi Horiguchi, Masakazu Aoki, Yoshinobu Nakagome, Shin'ichi Ikenaga, and Katsuhiro Shimohigashi. An experimental large-capacity semiconductor file memory using 16-levels/cell storage. *IEEE Journal of Solid-State Circuits*, pages 27–33, February 1988.

[40] James B. Burr et al. A 20 MHz Prime Factor DFT Processor. Technical report, Stanford University, September 1987.

[41] Robert R. Johnson. Multichip modules: Next-generation packages. *IEEE Spectrum*, pages 34–48, March 1990.

[42] Max Stanford Tomlinson Jr., Dennis J. Walker, and Massimo A. Sivilotti. A digital neural network architecture for VLSI. In *IJCNN International Joint Conference on Neural Networks*, pages II:545–550, 1990.

[43] Howard Kalter, John Barth, John Dilorenzo, Charles Drake, John Fifield, William Hovis, Gordon Kelley, Scott Lewis, John Nickel, Charles Stapper, and James Yankosky. A 50ns 16Mb DRAM with a 10ns data rate. In *IEEE International Solid-State Circuits Conference*, pages 232–233, 1990.

[44] Teuvo Kohonen. *Self-Organization and Associative Memory*. Springer-Verlag, 1984.

[45] S. Y. Kung and J. N. Hwang. Digital VLSI architectures for neural networks. In *IEEE International Symposium on Circuits and Systems*, pages 445–448, 1989.

[46] S.Y. Kung and J.N. Hwang. Parallel architectures for artificial neural nets. In *IEEE International Conference on Neural Networks*, pages II:165–172, 1988.

[47] S.Y. Kung and J.N. Hwang. Ring systolic designs for artificial neural nets. In *INNS First Annual Meeting*, pages 390–, 1988.

[48] Bang W. Lee, Ji-Chien Lee, and Bing J. Sheu. VLSI image processors using analog programmable synapses and neurons. In *IJCNN International Joint Conference on Neural Networks*, pages II:575–580, 1990.

[49] Weiping Li. *The Block Z transform and applications to digital signal processing using distributed arithmetic and the Modified Fermat Number transform*. PhD thesis, Stanford University, January 1988.

[50] Weiping Li and James B. Burr. Parallel multiplier accumulator using 4-2 adders. US patent pending, Application number 088,096, filing date August 21, 1987.

[51] Weiping Li and James B. Burr. An 80 MHz Multiply Accumulator. Technical report, Stanford University, September 1987.

[52] P. Mars and W. J. Poppelbaum. *Stochastic and deterministic averaging processors*. IEE, 1981.

[53] Carver Mead. Neuromorphic Electronic Systems. *Proceedings of the IEEE*, pages 1629–1636, 10 1990.

[54] Carver Mead and Lynn Conway. *Introduction to VLSI Systems*. Addison-Wesley, 1980.

[55] A. Moopenn, A.P. Thakoor, T. Duong, and S.K. Khanna. A neurocomputer based on an analog-digital hybrid architecture. In *IEEE First International Conference on Neural Networks*, pages III:479–486, 1987.

[56] Takayuki Morishita, Youichi Tamura, and Tatsuo Otsuki. A BiCMOS analog neural network with dynamically updated weights. In *IEEE International Solid-State Circuits Conference*, pages 142–143, 1990.

[57] Alan F. Murray and Anthony V. W. Smith. Asynchronous VLSI neural networks using pulse-stream arithmetic. *IEEE Journal of Solid-State Circuits*, 23(3):688–697, 1988.

[58] L. W. Nagel. SPICE2: A computer program to simulate semiconductor circuits. Technical report, University of California, Berkeley, May 9 1975. Memo ERL-M520.

[59] J. Ouali and G. Saucier. Silicon compiler for neuro-ASICs. In *IJCNN International Joint Conference on Neural Networks*, pages II:557–561, 1990.

[60] Carsten Peterson and Eric Hartman. Explorations of the mean field theory learning algorithm. In *Neural Networks*, volume 2, pages 475–494. Pergamon Press, 1989.

[61] Dean A. Pomerleau, George L. Gusciora, David S. Touretzky, and H.T. Kung. Neural network simulation at Warp speed: How we got 17 million connections per second. In *IEEE International Conference on Neural Networks*, pages II:143–150, 1988.

[62] Scott R. Powell and Paul M. Chau. Estimating power dissipation in VLSI signal processing chips: the PFA technique. In *VLSI Signal Processing IV*, pages 251–259, 1990.

[63] Jack Raffel, James Mann, Robert Berger, Antonio Soares, and Sheldon Gilbert. A generic architecture for wafer-scale neuromorphic systems. In *IEEE First International Conference on Neural Networks*, pages III:501–513, 1987.

[64] Mike Rudnick and Dan Hammerstrom. An interconnect structure for wafer scale neurocomputers. In David Touretzky, Geoffrey Hinton, and Terrence Sejnowski, editors, *Connectionist Models Summer School*, pages 498–512. Carnegie Mellon University, 1988.

[65] J.P. Sage, K. Thompson, and R.S. Withers. An artificial neural network integrated circuit based on MNOS/CCD principles. In *Neural Networks for Computing, Snowbird, Utah 1986*, pages 381–385. American Institute of Physics, 1986.

[66] Mark Santoro and Mark Horowitz. A pipelined 64X64b iterative array multiplier. In *IEEE International Solid-State Circuits Conference*, pages 35–36, February 1988.

[67] Mark Santoro and Mark Horowitz. SPIM: A pipelined 64X64-bit iterative multiplier. *IEEE Journal of Solid-State Circuits*, pages 487–493, April 1989.

[68] Mark R. Santoro. *Design and Clocking of VLSI Multipliers*. PhD thesis, Stanford University, October 1989.

[69] Terrance J. Sejnowski and Charles R. Rosenberg. NETtalk: A parallel network that learns to read aloud. Technical Report JHU/EECS-86/01, Johns Hopkins University, 1986.

[70] D. T. Shen and A. Weinberger. 4-2 carry-save adder implementation using send circuits. In *IBM Technical Disclosure Bulletin*, pages 3594–3597, February 1978.

[71] M. Sivilotti, M. Mahowald, and C. Mead. Realtime visual computations using analog CMOS processing arrays. In *Proceedings of the 1987 Stanford Conference*, 1987.

[72] Chris J. Terman. User's guide to NET, PRESIM, and RNL/NL. Technical Report VLSI 82-112, Massachusetts Institute of Technology, 1982.

[73] L. Waller. How MOSIS will slash the cost of IC prototyping. *Electronics*, 59(9), March 3 1986.

[74] Shlomo Waser and Michael J. Flynn. *Introduction to Arithmetic for Digital Systems Designers*. CBS College Publishing, 1982.

[75] Takumi Watanabe, Yoshi Sugiyama, Toshio Kondo, and Yoshihiro Kitamura. Neural network simulations on a massively parallel cellular array processor: AAP-2. In *IJCNN International Joint Conference on Neural Networks*, pages II:155–161, 1989.

[76] Carol Weiszmann, editor. *DARPA Neural Network Study, October 1987 - February 1988*. AFCEA International Press, 1988.

[77] Neil Weste and Kamran Eshraghian. *Principles of CMOS VLSI Design - A Systems Perspective*. Addison-Wesley, 1985.

[78] William Wike, David Van den Bout, and Thomas Miller III. The VLSI implementation of STONN. In *IJCNN International Joint Conference on Neural Networks*, pages II:593–598, 1990.

[79] Kazuo Yano, Toshiaki Yamanaka, Takashi Nishida, Masayoshi Saito, Katsuhiro Shimohigashi, and Akihiro Shimizu. A 3.8-ns CMOS 16x16-b multiplier using Complementary Pass-Transistor Logic. *IEEE Journal of Solid-State Circuits*, pages 388–394, April 1990.

[80] Moritoshi Yasunaga, Noboru Masuda, Mitsuo Asai, Minoru Yamada, Akira Masaki, and Yuzo Hirai. A wafer scale integration neural network utilizing completely digital circuits. In *IJCNN International Joint Conference on Neural Networks*, pages II:213–217, 1989.

[81] Moritoshi Yasunaga, Noboru Masuda, Masayoshi Yagyu, Mitsuo Asai, Minoru Yamada, and Akira Masaki. Design, fabrication and evaluation of a 5-inch wafer scale neural network LSI composed of 576 digital neurons. In *IJCNN International Joint Conference on Neural Networks*, pages II:527–535, 1990.

Appendix

The Tables of Contents for Volumes I, II, III, and IV of *Neural Networks* are listed below.

IMPLEMENTATIONS

7 ANALOG ELECTRONIC NEURAL NETWORKS FOR PATTERN RECOGNITION APPLICATIONS
Hans P. Graf, Lawrence D. Jackel, and John S. Denker

8 VLSI ARCHITECTURES FOR NEURAL NETWORKS
Joshua Alspector

Current VLSI Implementation Efforts
VLSI signal processing
Associative memory and pattern classification chips
VLSI learning systems
Other VLSI implementation efforts

Architecture of a Learning Chip
Overview of structure and operation
Neuron
Noise amplifier
Synapse
Control signals

Performance Evaluation of Network
XOR tests
Unsupervised learning

Discussion
Expanding the network with multiple chips
Neural interconnect
Applications of learning systems
Issues for further work
Future large-scale learning systems
Importance of the neural style

References

VOLUME II

CONCEPTS

APPLICATIONS

IMPLEMENTATIONS

CONCEPTS

Discussion

References

APPLICATIONS

IMPLEMENTATIONS

VOLUME IV

CONCEPTS

Error Correction, Fault Tolerance, Self Repair
Degrees of fault tolerance
Fault tolerance in neural nets
Acquisition of fault tolerance
Granularity of fault tolerance

Faults in Neural Networks
Types of component failures
Architectural considerations
Faults considered in this study

Discrete Classification Task
Task description
Network specifications
Regular training and testing
Training with faulty hidden units
Prolonged training with single failures
Discussion

Noise Immunity
Noisy characters
Training with faults
Training with noise
Training with noise and faults
Discussion

Analog Function Approximation
Fault tolerance for analog output
Domain width of basis functions
Task and network specifications
Regular training
Training with faults
Design of fault tolerant analog approximation networks
Summary

Hierarchical Nets
Classification tasks with a supervisor
Retraining subnetworks
Function approximation with a supervisor
Discussion

Summary and Conclusions

References

5 CONSTRAINED NEURAL NETWORKS FOR PATTERN RECOGNITION

Sara A. Solla and Yann le Cun

Introduction

Layered Neural Networks

APPLICATIONS

Fully Analog Pulse Stream Synapses
Time-modulation synapse
Pulse stream/transconductance multiplier synapses
Why pulse streams?

Future Work

Acknowledgements

References

11 RESOLVING CONSTRAINT SATISFACTION PROBLEMS WITH NEURAL NETWORKS: THE AUTOMATIC LOCAL ANNEALING SOLUTION
Jared Leinbach

Introduction

Constraint Satisfaction Problems
Neural network representations

The Local Maxima Problem

The Annealing Analogy to Processing
Physical basis
Annealing in networks

The Boltzmann Machine
Problems with the Boltzmann Machine

Automatic Local Annealing
Mechanics
Simulations
Results
How ALA works: an intuitive description
The flexibility of ALA

Directions for Future Research
Schema processing
Learning algorithms

References

INDEX